Humanitarian Crises, Intervention and Security

This book presents a new framework of analysis to assess natural and man-made disasters and humanitarian crises, and the feasibility of interventions in these complex emergencies.

The past half-century has witnessed a dramatic increase in such crises – such as in Haiti, Iraq and Sudan – and this volume aims to pioneer a theory-based, interdisciplinary framework that can assist students and practitioners in the field to acquire the skills and expertise necessary for evidence-based decision-making and programming in humanitarian action. It has four major objectives:

- to provide a tool for diagnosing and understanding complex emergencies, and build on the concepts of state security and human security to provide a 'Snap-Shot Analysis' of the status quo;
- to provide a tool for analyzing the causes of crises as well as the related stakeholder field;
- to provide a frame to structure and analyze the information required to evaluate, monitor and/or design interventions for different actors on a project and/or program level;
- to combine concepts used in the humanitarian field with underlying theory in a practically relevant way.

The book will be of much interest to students of humanitarian intervention, human security, peacebuilding, development studies, peace studies and IR in general.

Liesbet Heyse is assistant professor in organization sociology at the Department of Sociology/ICS, University of Groningen, the Netherlands. She is author/editor of two books.

Andrej Zwitter is professor of International Relations, University of Groningen, the Netherlands. He is author/editor of three books.

Rafael Wittek is professor of sociology, Department of Sociology, University of Groningen, the Netherlands. He is author/editor of two books.

Joost Herman is professor of globalization studies and humanitarian action, University of Groningen, the Netherlands. He is author/editor of three books.

Routledge Studies in Intervention and Statebuilding
Series Editor: David Chandler

Statebuilding and Intervention
Policies, practices and paradigms
Edited by David Chandler

Reintegration of Armed Groups After Conflict
Politics, violence and transition
Edited by Mats Berdal and David H. Ucko

Security, Development, and the Fragile State
Bridging the gap between theory and policy
David Carment, Stewart Prest, and Yiagadeesen Samy

Kosovo, Intervention and Statebuilding
The international community and the transition to independence
Edited by Aidan Hehir

Critical Perspectives on the Responsibility to Protect
Interrogating theory and practice
Edited by Philip Cunliffe

Statebuilding and Police Reform
The freedom of security
Barry J. Ryan

Violence in Post-Conflict Societies
Remarginalisation, remobilisers and relationships
Anders Themnér

Statebuilding in Afghanistan
Multinational contributions to reconstruction
Edited by Nik Hynek and Péter Marton

The International Community and Statebuilding
Getting its act together?
Edited by Patrice C. McMahon and Jon Western

Statebuilding and State-Formation
The political sociology of intervention
Edited by Berit Bliesemann de Guevara

Political Economy of Statebuilding
Power after peace
Edited by Mats Berdal and Dominik Zaum

New Agendas in Statebuilding
Hybridity, contingency and history
Edited by Robert Egnell and Peter Haldén

Mediation and Liberal Peacebuilding
Peace from the ashes of war?
Edited by Mikael Eriksson and Roland Kostić

Semantics of Statebuilding
Language, meanings and sovereignty
Edited by Nicolas Lemay-Hébert, Nicholas Onuf, Vojin Rakić and Petar Bojanić

Humanitarian Crises, Intervention and Security
A framework for evidence-based programming
Edited by Liesbet Heyse, Andrej Zwitter, Rafael Wittek and Joost Herman

Humanitarian Crises, Intervention and Security

A framework for evidence-based programming

Edited by
Liesbet Heyse, Andrej Zwitter,
Rafael Wittek and Joost Herman

Routledge
Taylor & Francis Group
LONDON AND NEW YORK

First published 2015
by Routledge
2 Park Square, Milton Park, Abingdon, Oxfordshire OX14 4RN

and by Routledge
711 Third Avenue, New York, NY 10017

First issued in paperback 2016

Routledge is an imprint of the Taylor & Francis Group, an informa business

British Library Cataloguing-in-Publication Data
A catalogue record for this book is available from the British Library

Library of Congress Cataloging-in-Publication Data
A catalog record has been requested for this book.

ISBN 13: 978-1-138-23662-2 (pbk)
ISBN 13: 978-0-415-83039-3 (hbk)

Typeset in Times New Roman
by Wearset Ltd, Boldon, Tyne and Wear

Contents

List of figures ix
List of tables xi
List of boxes xii
List of contributors xiii
Preface xv

1 **The need for evidence-based programming in humanitarian action** 1
LIESBET HEYSE, ANDREJ ZWITTER, RAFAEL WITTEK AND JOOST HERMAN

2 **Existing frameworks for humanitarian crisis analysis** 12
LIESBET HEYSE

3 **Context analysis and securitization** 29
ANDREJ ZWITTER AND JOOST HERMAN

4 **From theory to analysis: H-AID methodology** 43
RAFAEL WITTEK AND ANDREJ ZWITTER

5 **Conducting a Comprehensive Context Analysis (CCA)** 53
ANDREJ ZWITTER

6 **The political context** 63
CHRISTOPHER K. LAMONT

7 **The economic context** 70
FLEUR S. MULDER AND BARTJAN J.W. PENNINK

8 **The social-cultural context** 86
CECILE W.J. DE MILLIANO AND BARBARA BOUDEWIJNSE

9 The health and food context 99
RENSIA R. BAKKER

10 The environmental context 115
PETER D.M. WEESIE

11 From context analysis to intervention design 132
LIESBET HEYSE

12 Stakeholder analysis: towards feasible interventions 149
RAFAEL WITTEK

13 Monitoring, evaluation and learning in humanitarian organizations 171
CHAMUTAL AFEK-EITAM AND ADRIAAN FERF

Conclusion 189
LIESBET HEYSE, ANDREJ ZWITTER, RAFAEL WITTEK AND JOOST HERMAN

References 195
Index 213

Figures

1.1	Overview of the humanitarian analysis and intervention design framework	8
2.1	The selected frameworks and their relation to the immediate aftermath of a humanitarian crisis	13
2.2	The ideal balance of qualities of a framework of humanitarian crisis analysis	16
2.3	The progression of vulnerability	22
2.4	A visualization of the sustainable livelihoods framework	24
2.5	The five selected frameworks placed in the three dimensional space of core qualities	25
3.1	Levels of analysis in H-AID	37
5.1	Post-emergency mapping of the six context dimensions	57
5.2	CCA mapping before and after a hypothetical emergency	60
10.1	The many ways in which human well-being depends on ecosystem services	118
10.2	Interplay between disaster-struck/refugee populations, humanitarian assistance and the environment	124
11.1	The Context-Mechanism-Outcome approach in humanitarian crises	136
11.2	From prioritized intervention domains to thinkable and suitable food security interventions	143
11.3	Crucial mechanisms behind successful emergency seed aid interventions	144
11.4	Required context conditions for successful emergency seed aid interventions	145
11.5	Fictitious analysis of a CMO configuration applied to a particular crisis context	147
12.1	Graphical representation of the outcome continuum for issue 1 'size'	160
12.2	Stability analysis for issue 1 'size'	162
12.3	Negotiation landscape for issue 1 'size'	166
12.4	Relationship analysis for stakeholder 'government'	167

13.1 Data collection and analysis cycles used in the project management 177
13.2 5i framework for implementing evaluative practices 186
C.1 Overview of the humanitarian analysis and intervention design framework 191

Tables

1.1	The three core components of the H-AID framework	9
2.1	The five selected frameworks compared	26
4.1	Source reliability rating matrix	49
4.2	Data accuracy rating matrix	50
5.1	Contextual security levels and descriptors	54
5.2	Confidence intervals	56
5.3	Scale of security levels	57
5.4	Comprehensive context security levels after a hypothetical disaster event	58
5.5	Scaling of pre- and post-incident scaling	59
6.1	Key definitions	65
6.2	Political vulnerabilities in the Political Instability Index of the Economist Intelligence Unit	66
6.3	List of potential indicators for measuring the political dimension	67
6.4	Core groups and suggested methods	68
7.1	Core dimensions of the economic context	80
7.2	Additional dimensions of the economic context	82
8.1	Key concepts and diagnostic questions describing the socio-cultural context	96
8.2	Weighing enabling and disabling categories in the social-cultural context	98
9.1	Multiple levels of health security	110
9.2	Levels of food and nutrition security, ordered from national/regional to household/individual measurement	112
10.1	Major environmental disasters and their impact	123
12.1	Stakeholder description for issue 1: 'size of remote unit'	158
12.2	Summary of each stakeholder's position, salience, power for issue 1 ('size')	159
12.3	Stakeholder classification (from Mission's perspective)	163
13.1	Overview of monitoring and evaluation types and modalities	178
13.2	Qualitative and quantitative applied data collection methods	183

Box

11.1 Information sources about emergency food interventions 142

Contributors

Chamutal Afek-Eitam is currently finishing her PhD in organizational change, organizational learning and evaluative practices in humanitarian INGOs at the University of Groningen, the Netherlands. In addition, she is the CEO of the 3 Million Club, as well as co-managing the Humanitarian Genome and an associate trainer at REDR UK.

Rensia R. Bakker (Renée) works at the Institute of Medical Education at the Faculty of Medicine, University of Groningen, the Netherlands. She is coordinator of the Global Health profile in the International Bachelor of Medicine and teaches the Public Health module in the International Master of Humanitarian Action of the Network of Humanitarian Action (NOHA).

Barbara Boudewijnse is an anthroplogist and teaches at the University of Groningen, the Netherlands, in both the faculty of Theology and of Arts. She teaches the Anthropology module at the NOHA Master Program of Humanitarian Action at the University of Groningen, the Netherlands.

Cecile W.J. de Milliano is Programme Coordinator (Ad Interim) of Partners for Resilience (PfR) at CARE, Freelance Associate Trainer at REDR UK, and lecturer and researcher at the Globalization Studies Institute at the University of Groningen, the Netherlands. Her main areas of interest are resilience, disaster risk reduction, humanitarian action, climate change, child agency, child participation and gender equality.

Adriaan Ferf is an independent consultant and evaluator and guest lecturer at the NOHA Masters in Humanitarian Action at the University of Groningen, the Netherlands. He was team leader of numerous evaluation studies for a wide range of international organizations, among others, for the Netherlands and Belgium governments and international consortia in over 20 countries in Africa, Latin America and Asia.

Joost Herman is full professor of globalization studies and humanitarian action at the Faculty of Arts, University of Groningen, the Netherlands. He is also director of the Groningen University Globalization Studies Institute and program director of the Groningen Masters of Humanitarian Action of the Network of Humanitarian Action (NOHA), as well as financial director of the European NOHA association.

Liesbet Heyse is assistant professor of organization sociology at the Department of Sociology/ICS, University of Groningen, the Netherlands. Her research interests concern humanitarian aid provision, professionalization in the international nonprofit sector, and public sector reforms. She is co-managing the Humanitarian Genome and teaches in the NOHA Masters of Humanitarian Action at Groningen.

Christopher K. Lamont is assistant professor in the Department of International Relations and International Organization at the University of Groningen. He is also co-Chair of Research in Ethics and Globalization within the inter-faculty research institute Globalization Studies Groningen. He has published widely in the fields of transitional justice and human rights. His publications include his book *International Criminal Justice and the Politics of Compliance* (Ashgate 2010) and contributions to numerous journals and edited volumes.

Fleur S. Mulder was a student of the International Finance Management master (University of Groningen) at the moment of writing this chapter. This chapter also provided input for her Master's thesis which is based on empirical data collected at Dar es Salaam, Tanzania.

Bartjan J.W. Pennink is assistant professor at the Faculty of Business and Economics, University of Groningen, the Netherlands. He is especially interested in issues of local economic development in the non-Western world and teaches the Management module in the NOHA Masters of Humanitarian Action in Groningen.

Rafael Wittek is full professor of theoretical sociology at the Department of Sociology/ICS, University of Groningen, the Netherlands. His research interests are humanitarian action, social networks, the sociology of organizations and the sociology of aging. He has taught in NOHA for many years and currently teaches in the Humanitarian Management and Logistics executive master program of Lugano University.

Peter D.M. Weesie is an ecologist and environmentalist and works at the Science and Society Group of the Faculty of Mathematics and Natural Sciences, University of Groningen, the Netherlands. He is project leader of development cooperation projects with African universities in the fields of ecology, environment and food security. Peter Weesie also teaches on issues of environmental security in the NOHA Masters of Humanitarian Action in Groningen.

Andrej Zwitter is professor for international relations and head of the Political Science section at the Faculty of Law, University of Groningen, the Netherlands. He is also an honorary Senior Research Fellow at Liverpool Hope University, UK. He is coordinator of the second semester of the NOHA Masters in Humanitarian Action and teaches the module on context and stakeholder analysis therein.

Preface

In the past decade, the humanitarian sector has gone through – and still is going through – an impressive professionalization process. This shows, amongst other things, in the proliferation of humanitarian teaching and research programs all over the world, as well as sector-wide activities aimed at improving accountability and performance, such as the work done by ALNAP and the activities supported by the Humanitarian Innovation Fund. This book, on the one hand, reflects these developments and, on the other, aims to deliver a modest contribution to the further professionalization of the sector by means of the Humanitarian Analysis and Intervention Design Framework (H-AID framework).

The framework presented in this book has been the product of ten years of humanitarian action research and teaching at Groningen University, a member of the Network on Humanitarian Action (NOHA), as well as at the Humanitarian Management and Logistics executive master program of Lugano University. Over the years Groningen academics from a variety of disciplines have thought of ways to properly use and analyze information in humanitarian programming for the benefit of improving humanitarian effectiveness and impact. More specifically, the search has been for tools and methods to improve information collection, processing and analysis in designing humanitarian aid projects in the phases of needs and risk assessments, program planning, monitoring, evaluation and organizational learning. The result has been a concise model: the Humanitarian Analysis and Intervention Design Framework (H-AID framework).

The H-AID framework has been created on the basis of thorough analysis of existing methods and tools for the humanitarian sector. Weighing the pros and cons of existing models, the result has become a 'meta-model' for context analysis and evidence-based humanitarian programming. Beyond being a mere meta-model based on existing work, H-AID also incorporates a set of tools developed by the editors, which helps practitioners to improve their analytic and intervention design skills in humanitarian crises.

The path towards this book was initiated by collaborative work of Rafael Wittek and Joost Herman, which laid the fundament of a systematic approach to crisis analysis then called 'Comprehensive Security'. Their work was extended by Andrej Zwitter and Liesbet Heyse in the subsequent years. With the help of the whole team of researchers and teachers related to NOHA Groningen – as well as the constant and constructive feedback of our students (both young adults just entering the humanitarian sector and experienced practitioners) – this book is the current end result of our journey: the framework of analysis, which we call H-AID.

In the final stages of the book writing process, Lisa Reilly of the European Interagency Security Forum (EISF) provided invaluable advice strengthening elements of operational security and the practical side of disaster analysis and intervention design. We also thank Henk Hokse of *Graphique du Nord* for streamlining and formatting many of the figures in this book.

1 The need for evidence-based programming in humanitarian action

Liesbet Heyse, Andrej Zwitter, Rafael Wittek and Joost Herman

Ongoing challenges to provide impactful humanitarian aid

The field of humanitarian action continues to struggle with the challenge to conduct impactful humanitarian aid interventions. In part, this is because humanitarian action as a professional field cannot build on generalizable principles, large-scale datasets and case studies to provide one-size-fits-all intervention strategies. Rather, aid organizations have to design their interventions every time to fit the conditions and circumstances of an area with its cultural, political, economic, health and environmental specificities.

That these challenges persist, has been made painfully evident again with the Haiti earthquake. After the earthquake in 2010, Haiti experienced a massive influx of a variety of humanitarian and military actors. Due to the fact that the country's infrastructure was almost totally destroyed, the logistical side of humanitarian aid provision presented a tremendous challenge to the sector. However, this was not the only problem faced by humanitarian actors. Evaluations of the Haiti aid operations pointed at recurrent 'classic' flaws in the aid activities (CARE/Save the Children 2010, Patrick 2011). For example, many emergency products such as bottled water and medicines were imported and freely distributed while they were locally available. The influx of free goods in the local economy resulted in local businesses having to shut down, thereby further weakening the already weak economic infrastructure of the country. Evaluators attributed this mistake to a lack of contextual understanding (CARE/Save the Children 2010: 30).

Also, information was lacking with regard to targeted food distributions; there was no documentation that outlined community information strategies or any orientation for staff involved (CARE/Save the Children 2010). In addition, the quality of targeting and level of coverage was criticized. This problem was deemed to be due to a lack of available data and regular profiling of the population (Patrick 2011). All in all, as an OECD synthesis of evaluation reports summarized:

> Largely unfamiliar with humanitarian natural disasters in urban areas and compounded by poor contextual understanding of Haiti's society and economy and of the capacity of key stakeholders, the humanitarian

> community's reaction was a classical response: self contained, working outside government systems and reliant on imported material and personnel, supporting displaced individuals in internally displaced persons camps with food and non-food assistance.
>
> (Patrick 2011: 4)

The Haiti earthquake showed that the humanitarian sector is confronted with a persistent challenge to provide impactful aid. One important explanation for the persistence of this challenge is, as suggested by the above evaluators, the lack of analysis of humanitarian crisis situations and the use of these analyses in humanitarian work. This lack of understanding of the context can also hamper the design of appropriate risk management measures in humanitarian projects. In addition, lack of analysis and correct use of information can lead to 'classical mistakes' which seriously affect the outcomes of humanitarian aid operations and can even lead to more harm than good, as not only the Haiti earthquake but other humanitarian crises have shown. For example, also after the 2004 Indian Ocean earthquake, subsequent Tsunami evaluators stated that ineffective and inappropriate recovery programs were implemented due to a lack of awareness of the context at both the individual and institutional level (TEC 2006). This book focuses on ways to improve the use of information and analysis in humanitarian programming.

Humanitarian mission versus humanitarian reality

The core mission of humanitarian organizations is to save lives at risk and to relieve suffering of victims of 'man-made' or 'natural disasters' by providing high quality aid to those who need it most (Barnett and Weiss 2008). In its classic meaning, humanitarian aid is focused on the immediate aftermath of a crisis, conflict or disaster, which is then followed by phases of reconstruction, rehabilitation and development. However, given the increased acknowledgement of the structural causes of humanitarian crises – in terms of lack of preparedness, prevention and resilience (see also Chapter 2) – humanitarian aid projects also often have elements of prevention, risk reduction and 'building back better' (Fan 2013). In this book, we therefore adhere to this more extended interpretation of humanitarian aid.

Humanitarian aid requires the swift and correct identification and diagnosis of humanitarian problems and groups in need, next to decisions whether to start providing aid and, if so, whom to help, where, when and how. In addition, it requires a thorough risk assessment in order to assure that aid can be delivered safely. Once an aid project is started, it requires continuous monitoring of project activities, the context as well as the risk management strategies, in order to assure quality and safe access, and adapt activities when necessary. The humanitarian programming process in the sector is therefore often presented in terms of a project management cycle, referring to a step-wise process of identifying and formulating programming activities, planning and implementing activities, monitoring and evaluating them, as well as adapting activities when necessary (see for example, ECHO 2005 and ICRC 2011). These activities are employed with the aim – at the minimum – to prevent that no harm is done (Anderson 1999). The ambition of humanitarian aid providers is however much higher, namely to provide high quality aid that has impact, is appropriate, effective, sustainable and efficient, and meets all kinds of other quality criteria such as for example outlined in the Sphere standards.

In the past decades the sector has worked hard to get closer to meeting these ambitions. During and after the humanitarian operations in the 1990s (Rwanda, Somalia, Kosovo), the sector was criticized extensively for its lacking performance and the unintended, harmful consequences of its actions (Sommer 1994, Whitman and Pocock 1996). The sector was called to increase accountability and transparency, to professionalize its operations and to improve coordination (Edwards and Hulme 1996, Brown and Moore 2001, Choudhury and Ahmed 2002). This led to many initiatives, such as the Sphere project, the Cluster approach, the Good Humanitarian Donorship Initiative, the founding of ALNAP and ELHRA, and increased attention to Operational Security Management (HPN/ODI 2010). In addition, aid workers started to regularly follow training and obtained academic degrees in humanitarian action.[1] Furthermore, scholarly attention to the work and functioning of humanitarian organizations has grown tremendously.[2]

As a result of these initiatives, much has been learned about and done to improve the sector's performance, with positive results (de Waal 2010). To name a few: there is now increased recognition of the importance of involving local actors and staff in humanitarian programming, implementation and evaluation (see, for example, ALNAP 2003, Barry and Barham 2012), next to increased awareness of the importance of applying coherent and comprehensive approaches to humanitarian crises (Cahill 2007), as represented in the UN cluster approach, or attention to disaster risk reduction and resilience (Twigg 2004) and the link between relief, rehabilitation and development (International Review of the Red Cross 2011). Also, systematic approaches to security risk management for maintaining access have become mainstream, such as reflected in the *Saving Lives Together* recommendations of the IASC Steering Group on Security (2011) and the UN *To Stay and Deliver* document (OCHA 2011).

Nevertheless, the humanitarian mission to relieve suffering by providing high quality aid in an effective, appropriate and impactful way remains an ongoing challenge. This is partly due to the context in which humanitarian organizations' work is done, which is – maybe more than ever – characterized by complexity and constant and rapid change. This context presents humanitarian actors with multiple challenges in terms of access, security, political and cultural constraints, and issues of funding and staffing. For example, governmental donors increasingly view humanitarian assistance as part of a larger military strategy aimed at international stability and defense, which has created a complex relationship between humanitarian organizations and the military (ALNAP 2012a). Also, funding has become more dependent on these strategic interests. Furthermore, humanitarian aid workers continue to face risks of being wounded, murdered or taken hostage (see, for example, the Aid Worker Security Database at www.aidworkersecurity.org) whereas at the same time humanitarian organizations are frequently obstructed by political actors (ICRC 2012) or need to engage with a complex set of other actors, such as private development companies, security providers and infrastructure providers.

Nevertheless, the difficult context of aid provision is not the only cause for problematic aid provision. The organizations themselves also do not live up to the expectations at all times. As the Haiti example shows, humanitarian organizations make (repetitive) mistakes in their programming. These are partly deemed to be the result of standardized and supply-driven responses to humanitarian crisis and the lacking use of information and analysis (Bradt 2009, Darcy 2009). This has not only been the conclusion of evaluations of major humanitarian responses, such as after the Tsunami (2005) and the Haiti earthquake (2010), but is also subject of a steady stream of professional and academic publications (see, for example, Coyne 2013, Levine *et al.* 2011, Donahue and Tuohy 2007). The use of seeds and tools programs in humanitarian aid illustrates this point. Such programs can be a suitable response when targeted households do not have access to seeds and tools, when there is a lack of good quality seeds and tools, and this lack negatively impacts production (Maxwell *et al.* 2008). However, humanitarian seeds and tools programs are implemented in situations that do not meet these conditions (Levine and Chastre 2004), resulting in adverse effects, as has been painfully illustrated for the case of Caluquembe in Angola: seeds programs promoted maize and beans production, which resulted in a monoculture, inducing low prices on the market and heavy reliance on a single source of income, leaving households extremely vulnerable to shocks (Van Dijkhorst 2011). So, it is one problem if humanitarian aid is not as effective as one would hope it would be, but yet another if aid leads to harm and deterioration of people's lives.

The humanitarian sector is thus facing a continuous challenge to improve its performance and reduce (repetitive) mistakes in an ever increasing complex environment. This challenge needs to be addressed as part of the sector's duty to be accountable, not only to their donors and the general public, but also – and especially – to the aid recipients. This book aims to contribute to address this challenge by focusing on one particular aspect of humanitarian programming: tools and methods to improve information collection, processing and analysis in designing humanitarian aid projects in all phases of the project management cycle.

Using evidence in programming

Information collection, processing and analysis are key to successful and impactful humanitarian programming, but are often lacking in humanitarian programming (Maxwell *et al.* 2013a, Levine *et al.* 2011, Darcy 2009, Haan *et al.* 2005). Without information and the proper analysis of it, it is impossible to make informed decisions about humanitarian aid and how to deliver it safely. However, not all information is equally relevant. Particular types of information especially contribute to better humanitarian programming, as is nicely illustrated in the OECD Haiti evaluation synthesis:

> Largely missing from these assessments … were *contextual analyses* (particularly on political and economic issues) and *capacity assessments of Haitian stakeholders* (most notably the Haitian government) which would have allowed the humanitarian community a greater understanding of Haitian social and political dynamics and of the capacities of their natural Haitian partners across government and civil society to engage with and even lead recovery. Compounding these gaps in analysis, *valuable studies and assessments conducted by Haitians themselves were largely ignored.*
>
> (Patrick 2011: 3)

Hence, one way to address the persistent need for improved programming is by means of better, correct and increased use of relevant information in humanitarian planning and programming.

We specify the distinction between information and relevant information in the humanitarian sector with the term 'evidence' (Bradt 2009, ALNAP 2012b, Darcy and Knox 2013). This term originates from medical research in which evidence-based medicine is often practiced, referring to a highly specialized method based on quantitative methods, the use of baseline data and randomized controlled trials (RCTs) to test what medicines work for whom and how (Bradt 2009). In the humanitarian sector, the term 'evidence' is also used, but in a much broader and looser way, and for good reasons. One of such reasons is that it is not possible, or ethical, to work with control groups in humanitarian crises (not receiving any aid). Hence the impact of aid provision cannot be measured by means of RCTs. Second, both quantitative and qualitative methods can be useful and necessary to generate evidence. Moreover, in some crisis contexts only small-scale qualitative data collection might be possible, due to access or security problems. The meaning of 'evidence' for the humanitarian sector is therefore not as restricted to the definition used in medicine, as the following definition by ALNAP clearly shows: 'Evidence is true or credible information (quantitative and qualitative) that helps demonstrate the truth or falsehood of a given proposition' (ALNAP 2012b: 4).

In the humanitarian context, the term 'proposition', as mentioned in the above definition, refers to questions about (1) the existence of an actual or potential crisis, (2) 'what works' in preventing or mitigating crisis and (3) what is the most appropriate response (ALNAP 2012b, Darcy and Knox 2013). In other words, evidence in humanitarian action is required to answer two basic questions. First, a *diagnostic question*, referring to a thorough problem analysis, analyzing the symptoms of the crisis, the context in which it is taking place as well as the causes leading to the crisis. Which antecedents and mechanisms led to the crisis or emergency? How are these factors interrelated? Information from needs assessments, early warning exercises or situational reports often feed into this situational analysis (cf. Maxwell *et al.* 2013b, Darcy 2009). Second, there is the *intervention question* which refers to a thorough analysis of the proposed intervention. Given the diagnosis, what are suitable, feasible and safe options to intervene in a specific context? This requires a creative process – also referred to as response analysis (Maxwell *et al.* 2013a, 2013b, Darcy 2009) – in which one generates well-reasoned and substantiated arguments and evidence as to why certain interventions can be assumed to result in positive effects, both in the short and the long term. Evidence from the diagnostic analysis and previous interventions (evaluations, lessons learned exercises) is of importance in this reasoning process, next to information on the stakeholder field.

The depth and frequency with which one attempts to answer these questions may differ: in the initial emergency response phase this may have to be based on limited evidence, but as suggested in the project management cycle, program design and implementation should include continuous collection of more and up to date evidence that is used for monitoring purposes, so that projects and programs can be adjusted when more or other information becomes available.

These two questions of diagnosis and intervention will be at the heart of this book. We aim to assist the humanitarian sector with acquiring the analytical skills and tools in evidence production, collection, reduction, synthesis and analysis in order to answer these questions, as crucial part of the project management cycle and larger quality assurance processes. Put differently, this book focuses on the role of evidence and information during needs and risk assessments, program planning phases, monitoring activities and evaluation and organizational learning processes.

Obstacles to information collection, use and analysis

Whereas nobody in the humanitarian sector would deny the necessity of basing program decisions on information and analysis, the actual use of evidence in humanitarian programming remains limited (Darcy and Hoffman 2003, Darcy *et al.* 2007, Darcy 2009, Maxwell *et al.* 2013a). This is not so much caused by the fact that the sector lacks tools, methods and models to analyze humanitarian crisis for the purpose of programming. On the contrary, there are quite some analytical frameworks that can be used to diagnose humanitarian problems, such as the pressures and release model (Wisner *et al.* 1994/2005), the vulnerabilities and capacities model (Anderson and Woodrow 1989) and livelihood models (Chambers and Conway 1991, Morse *et al.* 2009). There are also sector-specific tools, such as the food security and nutrition response analysis tool (Maxwell *et al.* 2013a, 2013b; see also Darcy *et al.* 2013 for an overview).[3] Finally, there are tools especially for organizational and operational security management (HPN/ODI 2010). All these models, if implemented correctly, can provide a sound basis for programming decisions. Nevertheless, in many cases these models are not used for the design of aid projects. The reasons for this are manifold, to name a few (Bradt 2009, Darcy *et al.* 2013, Maxwell *et al.* 2013a):

- As stated previously, the sector works in a highly dynamic context characterized by constant change next to security and access issues. This complicates ambitions to swiftly obtain sufficient and reliable information for humanitarian programming.
- The sector works under time pressure: there is the imperative to act speedily in order to limit suffering as much as possible, especially nowadays since social media reports on humanitarian needs and suffering are almost instantaneous. This induces high expectations of the sector, with public and private donors wanting to see action quicker, which may result in an 'act then think' attitude, in which little time is taken to collect and process information before acting.
- There are gaps between what donors wish to fund, what organizations prefer to do and what the needs on the ground are, resulting in risks of mismatch between needs and aid provided. In the humanitarian sector, there can be strong preferences for certain types of aid, because they are more visible (in-kind food aid versus preventive action) or because these match with other goals (such as food market surpluses in the West).
- The sector is characterized by high turnover and lacking capacities for information processing, contributing to lack of institutional memory.

In such a context, the use of in-depth and rigorous information collection and analysis methods and tools is difficult, especially given the fact that speed in humanitarian action is required. Nevertheless, there is a call and need for using evidence for programming, risk management and quality assurance in a rigorous manner. Is there a way in which the above obstacles can (partly) be overcome, so that both speed and rigor can guide the generation and application of evidence in humanitarian programming? We claim this is possible by providing the sector with an analytical framework that is as lean as possible, without compromising quality standards of information collection and analysis.

The Humanitarian Analysis and Intervention Design Framework (H-AID)

In this book, we present to the sector a concise model to answer the above-mentioned diagnostic and intervention question: the Humanitarian Analysis and Intervention Design Framework (H-AID framework). This framework has been generated on the basis of a thorough analysis of advantages and disadvantages of existing methods and tools for the humanitarian sector (see Chapter 2). This analysis facilitated the translation of the essentials of existing models in an understandable and hands-on way into the H-AID framework, so that humanitarian staff can quickly learn and apply this in their work. In other words, the H-AID framework can be regarded as a 'meta-model' for context analysis and informed humanitarian programming because it synthesizes and combines core elements of existing frameworks. By doing so, the H-AID framework provides sufficient rigor for sound analysis, while at the same time taking the specific demands of humanitarian work into account. As stated by others (Maxwell *et al.* 2013a, 2013b, Darcy 2009), such a meta-model can function as a roadmap that guides practitioners to the necessary steps for thoughtful analysis and intervention in a comprehensive but concise way. The H-AID framework, however, goes beyond synthesizing other frameworks. It develops its own toolbox (Table 1.1).

In Chapter 2 we will argue how we arrived at the H-AID framework and its core components by comparing and evaluating existing analytical frameworks for humanitarian diagnostics and program design on various dimensions. Based on this analysis, we came to the H-AID framework that consists of the following core components, which are reflected in the structure of the book, namely: context, interventions and stakeholders. We develop specific tools that facilitate the analysis of crisis contexts, interventions and stakeholders (see Table 1.1).

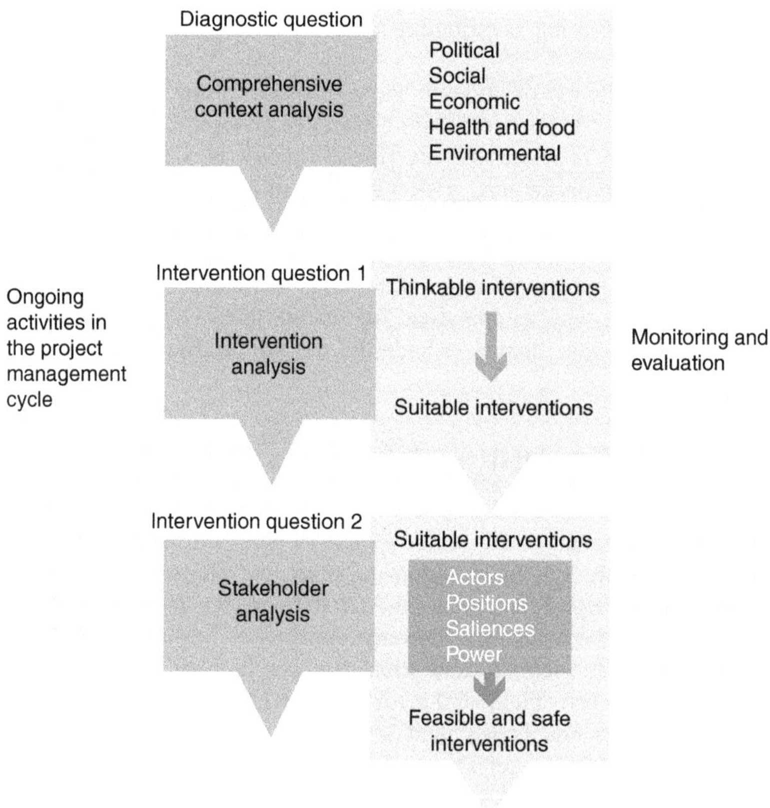

Figure 1.1 Overview of the humanitarian analysis and intervention design framework.

Table 1.1 The three core components of the H-AID framework

Core components of the framework	*Description*	*Chapters*
Comprehensive Context Analysis	Analysis of the social, political, economic, health, food and environmental context, preferably before and after a crisis, also related to risk assessment.	3–10
Intervention Analysis	Generating thinkable and suitable programming options and analyzing prior to the proposed project's start if, why and how the proposed activities will generate intended outcomes and effects, including safe access and safe aid provision, risk assessment and risk management (theory-based *ex ante* evaluation).	11
Stakeholder Analysis	Analyzing to what extent the generated suitable interventions can also be considered feasible and safe interventions given the stakeholder field one is operating in.	12
Monitoring and evaluation strategies	Defining and designing purposeful monitoring and evaluation tools; proper placing of these tools in the program/project design vis-à-vis organizational learning and accountability.	13

The basic thought behind these core components in our framework (context, interventions and stakeholder) is as follows (see also Figure 1.1). First, a thorough analysis of the context of a humanitarian crisis is required for a satisfactory answer to the diagnostic question: to what extent is there an actual or potential crisis and what is its nature? We propose to focus on those context dimensions prominent in existing frameworks for humanitarian needs assessments: the social, economic, political, health and food, and environmental context. In such an analysis one will also probably identify a wide range of actors that are of relevance in the particular emergency. Based on such a comprehensive context analysis, capacities and vulnerabilities in a particular humanitarian crisis can be identified with help of the *Comprehensive Context Analysis* (CCA) tool. This tool provides a quick overview of potential domains where intervention might be required.

Second, in the *Intervention Analysis* step we propose to use the generated context information for purposes of humanitarian programming by making a start to answer the intervention question by identifying potential domains for intervention. If a domain of intervention has been identified (such as food), we then propose to generate a set of thinkable options for action first (for example, within the domain of food security interventions there are multiple options such as free food distributions, cash transfers or seeds and tools programs, see also Maxwell *et al.* 2008, 2013a, 2013b). Based on this set of thinkable options, the method of *theory-based ex ante evaluation* is proposed to analyze which of these options are indeed suitable to the problem and context at hand. This method thus allows analysis prior to a project's start if, how and why a proposed intervention will generate the desired results.

The final step is to connect the identified set of suitable interventions to the stakeholder field in the crisis in order to identify which suitable interventions are also feasible and safe, given the goals, resources and relations of the other actors in the context as well as of the organization planning to intervene. This will also help to determine what risk management procedures are required to be able to maintain access and implement desired programs. By this we mean to actively manage risks, not simply avoiding them. The *Stakeholder Analysis* is thus in our view a very particular kind of actor analysis, in which it will be investigated – from the point of view of one focal actor (i.e. a humanitarian organization) that has identified a set of suitable interventions – what allies are available to pursue these interventions next to strategies to achieve the intervention's aims. The overall analysis allows to think over the impact the intervention may have on actors and power structures and what potential threats and harm the intervention may have to the organization and the community.

The framework and associated tools presented in this book have been developed and discussed as part of the second semester specialization of the Master of Humanitarian Action (NOHA) at the University of Groningen in the past ten years, and as a module in the Humanitarian Management and Logistics executive master program of Lugano University.[4] Ever since, these insights have been part of ongoing discussions with students, academics and practitioners.

The book aims to bridge theory and practice by providing humanitarian workers with academically founded, but practical tools, for sound context analysis and humanitarian programming. The book hence aims to enrich humanitarian programming practice by translating relevant insights from academia to the practical reality of humanitarian work. This bridging aim is reflected in the nature of the contributors to this book, which is a combination of academics in various disciplines and practitioners.

With this book, we hope to assist humanitarian aid workers to design impactful humanitarian interventions while at the same time to critically reflect on the way these interventions are planned and implemented. We do this by providing the sector with the essence of many analytical models for humanitarian diagnosis and intervention, and synthesize these into our model. We will also discuss the challenges of information collection, processing and analysis, and the use of evidence in humanitarian programming. Finally, we will discuss ways to connect insights gained from information collection and analysis to decisions about interventions.

This book is thereby targeted at all humanitarians interested in issues of information collection, analysis and use for purposes of context analysis, stakeholder analysis or intervention design. The book provides tools that can be used in, for example, internal trainings in humanitarian organizations and Master level courses in Humanitarian Assistance Programs, but also to those working in management, programming and evaluation positions in the sector.

Notes

1 See for example the Master programs offered by the Network of Humanitarian Action (NOHA) and Tufts University.
2 The increased academic attention to humanitarian action is illustrated by the founding of the International Humanitarian Studies Association in 2010 that has organized an annual meeting ever since.
3 Some of these models will more extensively be discussed in the next chapter.
4 This course was first titled 'Comprehensive Security in Humanitarian Action' and later 'Disaster Analysis and Intervention Design'.

2 Existing frameworks for humanitarian crisis analysis

Liesbet Heyse

Towards an encompassing but lean framework for humanitarian crisis analysis

This book aims to assist humanitarian aid workers with context, actor and intervention analysis by facilitating them with acquiring the analytical skills and tools in evidence production, collection, reduction, synthesis and analysis in order to plan for and design safe humanitarian interventions. However, it could be argued that already many tools and frameworks in this field exist that aim to do the same. Why then present yet another framework?

As stated in Chapter 1, our framework is a concise model generated on the basis of the thorough analysis of existing methods and tools. This model – called the Humanitarian Analysis and Intervention Design Framework (hereafter, the H-AID framework) – synthesizes and combines the essential elements of existing frameworks in an understandable and hands-on way so that humanitarian staff can quickly learn and apply this in their work. Hence, through this synthesis and combination we present a 'meta' model for evidence-based humanitarian programming that provides sufficient rigor for sound analysis, while at the same time taking the specifics of humanitarian work into account.

In this chapter, we present the core components of this framework and how we arrived at these. We therefore discuss the characteristics of a number of frameworks for analyzing humanitarian crises that have been extensively used and referred to in the humanitarian sector. The selected frameworks are: the Pressures and Release model (PAR), the Political Economy/Arena Approach (PolEc), the Multi-Cluster/Sector Initial Rapid Assessment (MIRA), Capabilities and Vulnerabilities Analysis (CVA) and the (Sustainable) Livelihoods Framework (SLF). Whenever appropriate, we will also discuss how these frameworks provide input in issues of organizational security management.

We opted for this particular selection of frameworks, since these frameworks offer what we would call a 'structural approach' to humanitarian crises. By this we mean that humanitarian crises are not perceived as sudden events, but as processes that build up and can have many manifestations throughout time. Although the distinction between natural and man-made disasters still is often made, the common wisdom nowadays is that humanitarian crises do not just happen like that, but are the consequence of high vulnerability of people to (disruptions in) their environment. Hence, the impact of a disruption – be it physical violence, an earthquake, drought or something else – is related to lack of (human) preparedness, prevention and resilience in society. Consequently, humanitarian aid is more than providing immediate relief; it is about connecting to existing or remaining capacities of individuals, groups, organizations and societies, and intervening in such a way that people's vulnerabilities are reduced, not only in the short term but also in the longer term. Put differently, humanitarian aid aims at the minimum to support existing capacities and resilience and at the most to contribute to build capacity, reconstruct societies and create resilience for the future.

The selected frameworks recognize this structural dimension of humanitarian crises, as will become visible in the remainder of this chapter, by emphasizing the importance of analyzing the *context* and (immediate) *causes* of humanitarian crises as well as the role of *actors*, three elements we deem crucial for this more structural approach to humanitarian crises. Also in the Good Practice Review on operational security management these three elements are – not coincidentally – crucial for achieving safe humanitarian interventions. In addition, the selected approaches and frameworks together cover a wide range of phases and activities that humanitarian organizations are involved in (see Figure 2.1): some approaches focus more on the immediate aftermath of a crisis (MIRA), whereas others are more distant from the immediate aftermath of a crisis, such as the CVA and PolEc approach or the PAR model and SLFs.

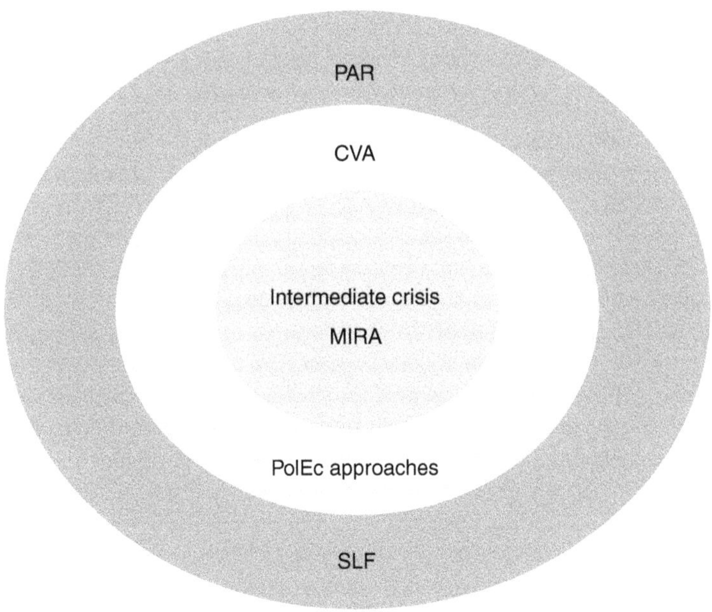

Figure 2.1 The selected frameworks and their relation to the immediate aftermath of a humanitarian crisis.

We evaluate these frameworks on a number of core qualities. In the next section, we outline the set of criteria developed to evaluate the selected frameworks on their usefulness, applicability and rigor for evidence-based programming in humanitarian action. In the sections that follow, we evaluate the selected frameworks on these dimensions, followed by a discussion on the differences and similarities of these frameworks. Based on this analysis, we present the core elements of the H-AID framework.

It should be noted that we by no means intend to present a complete picture of the selected frameworks; we aim to limit our discussion of these frameworks in relation to the core qualities discussed. For in-depth descriptions, reviews and discussions of each framework, we refer to the references listed with each framework.

Balancing core qualities of humanitarian crisis analysis frameworks

In order for any analytical framework to be used in the humanitarian sector, it needs to deliver and balance a variety of qualities. Below we outline three types of qualities that in our view need to be balanced in order for an analytical framework to be useful and used in the humanitarian sector. We contend these qualities to be of relevance because they help humanitarian workers and organizations to adhere to the humanitarian principles of humanity, impartiality and neutrality. The humanitarian principles are served in that a framework based on these core qualities will help aid workers to plan for aid on the basis of need.

Quick and 'dirty' analysis versus slow and thorough analysis

First of all, humanitarian work needs to be done quickly. Time and manpower in humanitarian crises is often limited, so any analytical framework for humanitarian crises should allow the speedy generation of results. The requirement for speedy results implies that the framework is not too extensive in terms of the elements that need to be analyzed and the amount of information that needs to be collected.

However, speed should not go at the expense of the quality of the information collected. It is crucial that any analytical framework for informed humanitarian programming provides the user with guidelines as to what are the relevant dimensions to focus on – whether these are context dimensions, actors or other dimensions – and how to generate reliable and valid information on these dimensions (see Chapter 4 for an elaboration of these concepts). In short, information is most likely to be reliable if it is transparently reported how data was collected and analyzed, so that the data collection and analysis process can be replicated and checked upon by others. Optimal reliability is achieved if this replication exercise leads to the same results, that is, when more analysts using the same data and techniques independently reach the same conclusions. Information and evidence can be called valid if the information generated is relevant to the situation at hand and correctly represents the dimensions of study in a particular context.

Based on the above, we distinguish between frameworks to the degree that they can be regarded as 'quick and dirty', in that they provide coarse-grained means to very quickly analyze a humanitarian situation by presenting quite a broad (unspecified) set of dimensions and indicators to focus on from which one can pick and choose, or as 'slow and thorough', meaning that these frameworks offer an extensive set of specified dimensions, indicators and interconnections to focus on which takes quite some time to collect information on. In other words, frameworks that offer a quick analysis of humanitarian settings often do this at the expense of the reliability and validity of the data gathered. 'Slow and thorough' frameworks offer high reliability and validity, but at the expense of speed. However, since it might be theoretically possible that frameworks can be both 'quick and thorough' or 'slow and dirty' we analyze each framework on both the time it costs for data collection and analysis, and the degree to which the framework is thorough, i.e. it specifies the dimensions and indicators to be studied.

General applicability versus specific applicability

Ideally, a framework for humanitarian crisis analysis can be applied to all phases of humanitarian crises – such as the immediate onset of a crisis or the recovery and rehabilitation phase – as well as to a variety of humanitarian crisis types, such as earthquakes, floods, conflicts or complex emergencies. This requires the framework to be so general that it allows applicability to a wide variety of crisis phases and contexts, while at the same it should ideally also be possible to adjust this general framework to specifics of particular crisis contexts. This general applicability can be achieved if the framework outlines a set of core components, possible indicators and methods as how to collect data on these components, whereas at the same time the framework also provides guidelines as to when and how to restrict, extend and adapt these core components.

'Just' an analytical tool versus providing guidelines for programming

A final quality of humanitarian crisis analysis frameworks is that it should assist humanitarian aid workers with making decisions on the basis of the evidence collected. This requires that the framework provides users with advice as how to interpret the evidence collected and how to translate this information into humanitarian interventions, activities and projects. Take for example considerations regarding a food intervention. Let us assume that one has collected reliable and valid evidence that food access is indeed problematic. The next question is then how to address this problem in a way that is likely to generate the desired effect (i.e. improved food access) and at the same time prevent the occurrence of undesired effects (i.e. distorted markets). Given the plethora of livelihoods and food interventions available – such as food for work, cash transfers or free food distribution – how is it decided which intervention is the best option in a given context? Ideally, a humanitarian crisis analysis framework should provide tools and advice as how to arrive at such decisions.

Balancing core qualities

Ideally, any humanitarian crisis analysis framework would balance these core qualities, being swift, adaptable, thorough and practically related to humanitarian intervention design by offering programming advice (see Figure 2.2).

One could question whether it is possible to arrive at such a balance. For example, at one extreme there might be quite extensive, detailed and rigorous frameworks for analysis, generating in-depth and highly reliable and valid evidence for specific crisis contexts. However, this might hamper the speed of applying such a framework as well as the adaptability to which it can be applied to different crisis contexts. At another extreme there might be quite general frameworks that focus on broad dimensions and a limited set of indicators. This might allow for flexibility and speedy analysis, but might not meet the qualities of relevant, reliable and valid evidence generation.

In the next sections, the selected frameworks for humanitarian crisis analysis are evaluated on the core qualities outlined in this section. We will discuss per framework its basic components and then proceed to evaluate these frameworks on the qualities set out above. We discuss the similarities and differences between the frameworks, and relate them to issues of organizational security analysis when appropriate, and then, based on this evaluation, we proceed with presenting our meta-framework and its core components.

The MIRA approach

The Multi-Cluster/Sector Initial Rapid Assessment is a method developed by the Interagency Standing Committee (IASC) as a rapid needs assessment tool to

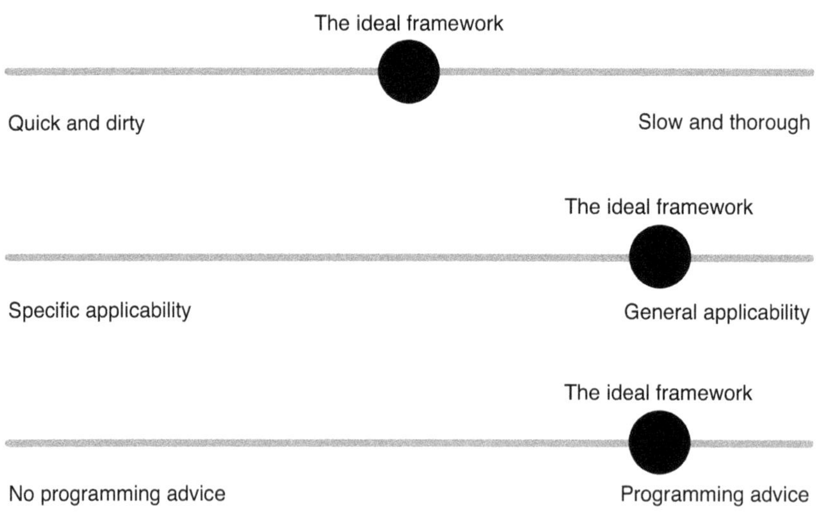

Figure 2.2 The ideal balance of qualities of a framework of humanitarian crisis analysis.

identify strategic humanitarian priorities during the first weeks of a humanitarian emergency (IASC 2012b: 3). It is explicitly meant as a method to be applied by all key stakeholders in a humanitarian emergency as a way to develop a common, cross-sectorial and multi-cluster understanding of the emergency. MIRA both requires secondary data analysis and community level assessments as a form of primary data collection. The secondary data analysis is meant to develop a so-called Preliminary Scenario Definition within 72 hours as a way to inform response planning and funding appeals. It consists of both a pre-crisis and an in-crisis analysis. The primary data analysis is meant for a more extensive rapid assessment of the situation by means of key informant interviews and observation, and should be ready within two weeks. For the key informant interviews a pre-structured interview scheme is provided, asking into issues of food, health, income, shelter, security, education, etc. Also, specific advice is given how to conduct such interviews. The same is provided for applying the method of observation.

In general, the approach focuses on identifying (IASC 2012b):

- the impact of the crisis in terms of drivers of the crisis (causal analysis), the scope of the crisis and the humanitarian profile (i.e. the number of people affected) and the status of the populations in the affected areas;
- the response capacity, in terms of national and international capacities and responses including local coping capacities;
- access and gaps, in terms of humanitarian access (logistically but also security wise), coverage and gaps in aid provision.

For each of these dimensions research questions are formulated, related to the status of certain groups and issues (such as protection), the impact of the crisis on these groups and issues, as well as the associated vulnerabilities, risks and trends. Examples of research questions are:

- What are the main drivers of the crisis (including environmental, socio-political, climatic and economic factors)?
- What are the known coping mechanisms of local communities and how were they affected?
- What are the main considerations affecting the local population and the delivery of assistance (armed groups, gender-based violence, sexual exploitation and abuse, and UXOs) and where are they?

The MIRA approach also pays attention to security issues by asking about the main security risks and civil–military relations, but not extensively.

When we relate this approach to the three core qualities, we label this approach as quite ‘quick and dirty’ in that it helps to quickly get an overview of the needs of groups in certain crisis-struck areas. It thus does not provide tools for an in-depth analysis. For example, there is no explicit causal model underlying the identification of crisis drivers (such as for example in the PAR model) and the dimensions mentioned above are not specified in detailed indicators, although for the primary data collection specification is provided. Moreover, the approach lacks the broader contextual analysis as proposed by other models, which makes sense given its aim to provide a quick overview. The approach is generally applicable to various crisis contexts but is especially meant for the immediate crisis phase, not as a prevention tool or as a means to make the transition from relief to development. In addition, the method is meant to help define strategic humanitarian vulnerabilities, not to assist agencies in making detailed decisions on localized responses and projects (IASC 2012b: 3). Hence, the approach offers programming advice, but quite general on a strategic level.

Political economy and arena approaches

A political economy approach to humanitarian crises explicitly relates the dynamics of humanitarian crises to the actors in a crisis and their interests, goals, resources and interaction with each other (Collinson 2002). This approach has as a central question whether groups and actors have something to gain or lose in humanitarian crises, and, if so, what exactly. Societal and crisis processes are thus viewed as inherently political, related to issues of power and wealth distribution. For example, violence is perceived as an opportunity for some groups in society to gain power and ensure access to resources, thereby leading to an interest in the continuation of conflict (Anderson 1999). Conflicts thus create their own economies, such as a war economy, a shadow economy and a 'survival' (or, coping) economy with distinct actors pursuing different goals. This approach is related to literature on 'greed and grievance' in conflicts (Collier and Hoeffler 2004).

In this approach vulnerability is thus very much conceptualized in terms of powerlessness. The approach stresses the importance of analyzing the actors in a crisis in order to achieve a good understanding of why some groups are less powerful than others. It also asks for a careful *ex ante* evaluation of the ways in which humanitarian aid could become a part of the various economies and what potential undesired consequences this could generate. It therefore sees humanitarian action and intervention as inherently political.

We also see the work by Hilhorst *et al.* as part of this approach, since it views humanitarian space as an arena in which various actors negotiate the outcomes of aid (Hilhorst and Serrano 2010, Hilhorst and Jansen 2010). The outcomes are the result of how actors involved (donors, recipients, governments) interpret the context, needs, their own role and each other. Actors are seen as driven by different motives, political and organizational ones included (Hilhorst and Jansen 2010). This so-called 'actor-oriented approach' differs slightly from a pure political economy approach, since it asks attention not only for power issues and relationships in humanitarian crisis, but also to how actors define and interpret a humanitarian crisis context. Nevertheless, both approaches emphasize the importance of analyzing the actors in a crisis and their goals, motives and resources as a way to understand how humanitarian crises develop and persist, and how humanitarian aid is part of these dynamics.

When we relate these approaches to the three core qualities defined previously, this type of approach first requires quite a thorough analysis of all actors in a crisis, and their perceptions, interests and motives. Especially the work by Hilhorst and colleagues is based on thorough anthropological fieldwork and in-depth understanding of local crisis contexts. Nevertheless, a more 'quick and dirty' version of this approach is possible by means of a more or less rough stakeholder analysis. Collinson (2002), for example, provides suggestions on what concrete dimensions to focus on and what methods to employ when applying this approach. We therefore rate it as a medium slow and thorough approach. Both approaches are also quite generally applicable. For example, whereas the political economy approach was first predominantly applied to conflict settings, it has also been applied to understand the differential impact of natural disasters on groups in society (see for example, Kenny 2009, Cohen and Werker 2008). Finally, in terms of programming advice, the approaches give opportunities to identify marginalized groups in society as potential target groups for aid and to decide on appropriate ways to help address the causes of their marginalization. The approaches also provide options to think through the impact of humanitarian aid on the dynamics of conflict. Moreover, such an approach resembles very much the actor analysis for purposes of organizational risk and safety management as recommended in the HPN Good Practice Review (2010).

The vulnerabilities and capacities framework

In 1989, Mary Anderson and Peter Woodrow introduced the idea of vulnerability and capacity analysis in their book *Rising from the Ashes.* This type of analysis aims to assist humanitarian aid workers in planning and implementing emergency relief programs whereas at the same time fundamental sustainable development is promoted (Anderson and Woodrow 1990: 7). It is therefore also called a 'developmental relief approach'.

The basic point of departure is not to view people in need as helpless victims, but as 'active, capable and inventive managers of their own lives' (Anderson and Woodrow 1990: 7). By assisting people to increase their capacities they can be empowered to lead the lives they wish to live, socially, economically, morally, etc. At the same time organizations can assist in decreasing these people's vulnerabilities to events that threaten their existence. Both capacities and vulnerabilities can be subdivided into various dimensions. These are (Anderson and Woodrow 1990: 10):

- Physical and material vulnerabilities/capacities, related to productive resources, skills and hazards, such as land, climate, health, infrastructure, labor and lack of income or food kept in storage.
- Social and organizational vulnerabilities/capacities, related to the relations and organization among people, such as formal political structures and informal social structures, next to issues of marginalization and lack of group solidarity versus social cohesion and community assistance.

- Attitudinal and motivational vulnerabilities/capacities, related to how the community views its ability to create change, which includes ideologies, beliefs, motivations and experiences of collaboration, such as low confidence and being apathetic versus a strong will to survive and motivation to recover.

If vulnerabilities are high and capacities low, then communities or societies are more prone to disasters and crisis. Communities can thus be helped to become more 'disaster resistant' by decreasing their vulnerability and fostering their capacities. Vulnerabilities differ from needs, in that they pinpoint at the deeper roots of the needs of people in crisis.

The three dimensions are quite broad and although examples are provided for each dimension, no extensive outline of their components and potential indicators is given in the original documents. Furthermore, it is advised to disaggregate the information on vulnerabilities and capacities by, for example, gender, class, religion, ethnicity, age. It can also be applied to different time points – to compare developments in capacities and vulnerabilities – and different levels (household, community, provincial levels, etc.). In later elaborations of the model, guiding questions for the analysis have been formulated (see, for example, CARE's climate vulnerability and capacity analysis (2009)). Also concrete tools for participatory analysis and checklists of vulnerabilities are suggested (Davis *et al.* 2004, CARE 2009).

The framework is explicitly presented as a diagnostic tool (Anderson and Woodrow 1990: 77) that is not aimed at providing guidance for specific actions or projects. It is claimed, however, that the framework offers ways to organize and systematize information that can help to identify possibilities for program responses. For example, it can help organizations to identify groups and organizations with which it might make sense to establish partnerships (see for example, ICRC 2007a). However, that this can be difficult is made clear in Heijmans and Victoria's work regarding the Philippine case:

> Project staff still find it difficult to apply the CVA as an analytical tool … the result is often more descriptive than analytical.… Members find it difficult to use the CVA for the identification of appropriate interventions.… Its use is limited to counter-check selected interventions.
>
> (2001:75)

Based on the above, it can be concluded that CVA can be rather quickly applied and can be regarded as medium thorough, in that the dimensions provided were originally not specified much, but later more specified indicators, checklists and questions were proposed (ICRC 2007a, 2007b, Heijmans and Victoria 2001, Provention 2007). At the same time, checklists are criticized (Davis *et al.* 2004). CVA also is quite generally applicable, in that it can also be applied as a prevention tool, although in terms of crisis contexts it is most used for disaster prone areas. Finally, CVA offers basic programming advice, in that it offers insights in potential vulnerabilities that one could strive to diminish as well as capacities that one could support, which helps to identify interesting options for local partnerships.

The Pressure and Release model (PAR)

The Pressure and Release model, developed by Wisner and Blaikie (1994/2005), is closely related to Anderson and Woodrow's pledge for thinking in capacities and vulnerabilities. The authors use this way of thinking to understand the risks of disasters as well as the effects of hazards by elaborating a model that sketches 'the progression of vulnerability'. Core concepts in this framework are risks, hazards and vulnerabilities, in combination with root causes, dynamic pressures and unsafe conditions (see Figure 2.3).

Risk is defined as the combination of potential hazards that threaten communities and the vulnerabilities of certain groups, meaning the likelihood that these hazards will actually harm groups and communities. Hence, the more vulnerable groups are, the higher the risk to be harmed by hazards. This framework thus points out that some societies and groups are more at risk for disasters and crisis due to certain hazards (i.e. natural disaster events, such as earthquakes or floods) and human vulnerabilities to such hazards. These vulnerabilities are explained in three intermediate steps. First, there is a set of *root causes* that form the most distant set of explanations of societal vulnerability. These root causes are related to social and economic structures, ideologies and history and culture. These root causes translate into a set of so-called *dynamic pressures* at the political and institutional macro level, leading to particular forms of insecurity that arise as a result of societal deficiencies – such as failing or lacking local institutions or lack of media freedom – as well as macro forces, i.e. rapid population growth, urbanization, conflict, economic crisis or declining biodiversity (Wisner *et al.* 2012). These pressures are called 'dynamic', because they 'transmit the historic weight of root causes along the "chain of causation", as an intermediary between them and fragile livelihoods and unsafe locations and conditions' (Wisner *et al.* 2012: 25). Third, the dynamic pressures result in a set of *unsafe conditions* related to the natural, physical, economic, social and political environment. These conditions are the specific ways in which a group's or community's vulnerability become visible in a certain place and time. This part of the model has later been relabeled as 'fragile livelihoods and unsafe locations' which is then strongly related to lack of access to resources and processes of marginalization (Wisner *et al.* 2012: 24). Together the root causes thus lead to dynamic pressures that in turn result in unsafe conditions that create vulnerabilities to disaster.

The authors also show how the progression of vulnerability could hypothetically be turned around into a process of progression of safety (see Figure 2.3). The model allows the identification of so-called entry points of intervention in terms of creating safer conditions, minimizing dynamic pressures or addressing root causes. For example, one could work towards reducing the likelihood that hazards occur by implementing disaster risk reduction measures, such as dams,

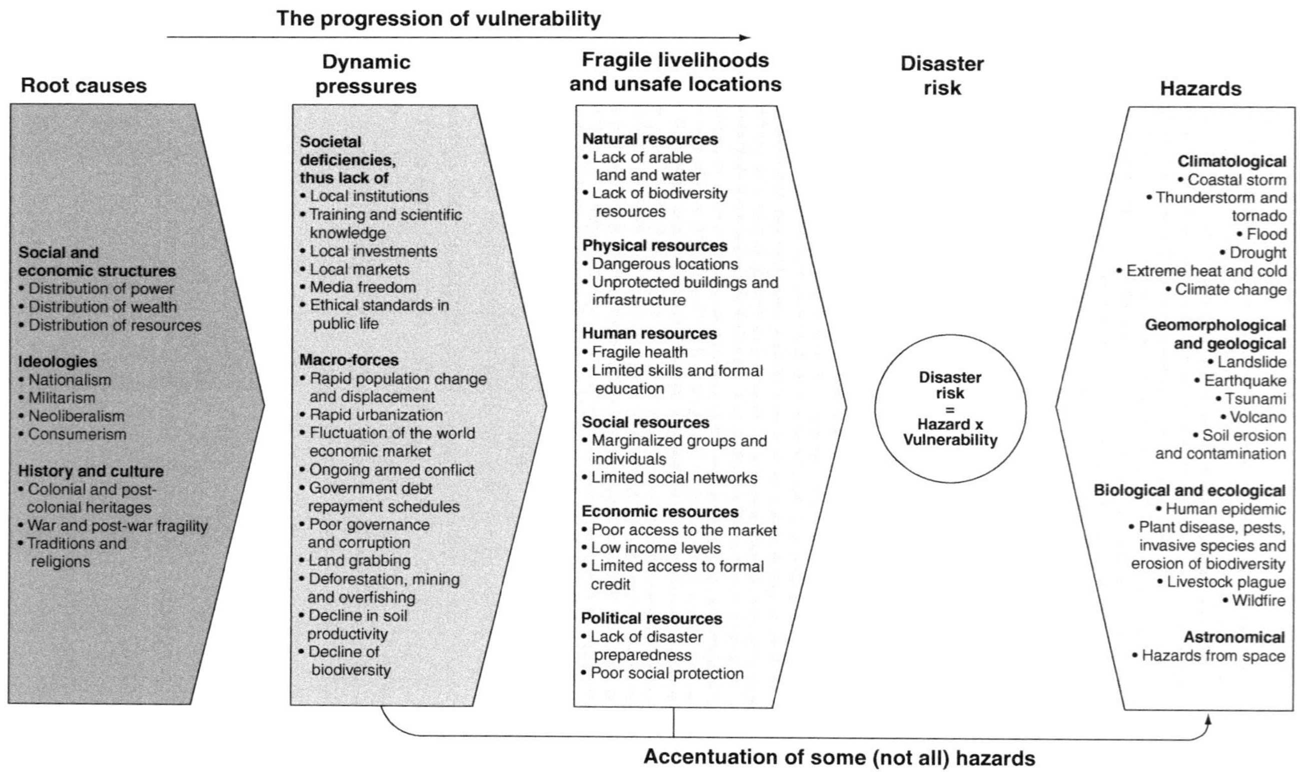

Figure 2.3 The progression of vulnerability (source: Wisner *et al*. 2012: 23).

shelters or monitoring systems. One could also try to influence the dynamic pressures by providing training or supporting institutions. Finally one could attempt to impact on root causes by addressing lack of access to power, for example.

Related to the three core qualities presented previously, one can see from Figure 2.3, the Pressure and Release model is quite extensive, providing a detailed overview and elaboration of potential factors that can explain vulnerability to disaster. In the *Handbook of Hazards and Disaster Risk Reduction* (2012), the full model is explained and elaborated, and for each hazard chapters are provided that define the particular hazard and ways to monitor and address these. There is also a historical dimension to this framework, which asks for in-depth and longitudinal research. It can therefore be categorized as a 'slow and thorough' framework for analysis. Furthermore, the model is specifically applicable in two ways. In terms of crisis phases, it is especially meant as a tool for prevention and disaster risk reduction, not for the immediate relief phase, since the model is so extensive and detailed. In terms of crisis contexts, the model is especially meant to be applicable to natural hazards, such as climatological, geomorphological/geological, biological/ecological and astronomical events. Finally, in terms of programming advice, the model offers various 'entry points of intervention' in the release model, due to its specific elaboration of dimensions in terms of root causes, dynamic pressures and unsafe conditions.

Livelihoods approach

A fifth relevant framework was developed as part of poverty reduction strategies, but also proves useful for understanding humanitarian crisis contexts (Twigg 2001). This framework focuses on the livelihoods of individuals and households as a way to understand better why some groups are more vulnerable (to hazards, for example) than others. A livelihood comprises the capabilities, assets (stores, resources, claims and access) and activities required for a certain level and means of living. A livelihood is sustainable when it can cope with and recover from stress and shocks, maintain or enhance its capabilities and assets, and provide sustainable livelihood opportunities for a next generation; and contributes net benefits to other livelihoods at the local and global levels and in the short and long term (Chambers and Conway 1991: 6, see Figure 2.4).

The aim of the framework is to uncover what factors and processes can help to improve the livelihoods of people, and thereby diminish their vulnerability to shocks and stress. The focus is on the following core elements: livelihood assets, transforming structures and processes, livelihood strategies and the vulnerability context (see Figure 2.4). The *livelihood assets* consist of the social, human, natural, physical and financial capital of individuals, households and groups. Human capital refers to the skills, knowledge and abilities people have, whereas social capital refers to the social support (in terms of relationships, networks and group membership) that people have. Natural capital points to the natural sources that livelihoods depend on (such as land, forests, etc.) and physical capital is about the basic infrastructure and producer goods needed to secure livelihoods.

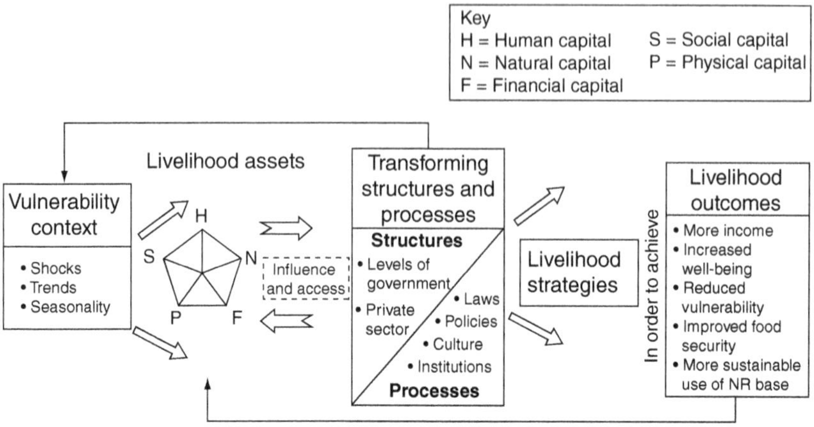

Figure 2.4 A visualization of the sustainable livelihoods framework (source: DFID 2001: 3).

The *transforming structures and processes* are the institutions, organizations (in the public, private and nonprofit sector), policies, norms and legislation that influence livelihoods in terms of access to and exchange between the various forms of capital. In addition, groups have all kinds of *strategies* to secure their livelihoods, such as seasonal labor, selling produce on markets, and diversified agriculture. By analyzing the factors that make up the livelihoods of communities and households, aid organizations can detect and support the positive aspects and at the same time try to diminish vulnerabilities. This is very much dependent on *the vulnerability context* that consists of contextual trends and shocks that influence livelihoods. Based on the analysis of the above, the SLF should provide guidance for intervention.

As one can see from the above and Figure 2.4, a livelihoods analysis, if done properly, consists of an in-depth study of all these factors. This makes the model quite time consuming to apply (Morse *et al.* 2009, Krantz 2001). The model is quite specified, and could thus allow for thorough analysis. Guidelines as how to analyze each dimension in the model have been provided for (especially by DFID 2001). However, there has been some criticism as to how exactly to study the dimensions in the model (Morse *et al.* 2009), for example as it comes to the assets and the transforming structures and processes. It therefore requires well-trained and highly skilled staff to conduct SLF properly (Carney 2002). We therefore label the model as slow and medium thorough. The model is quite generally applicable, also to for example conflict-ridden contexts, which is then often combined with a political economy approach (Collinson 2002, 2003, Lautze and Raven-Roberts 2006). In terms of programming advice, the model is quite specific so that it potentially offers various entry points for intervention. However, it has been argued that the model is so complex and that many factors in the model are difficult to influence by aid agencies that this limits the options for programming (Morse *et al.* 2009).

Comparing and combining the evaluated frameworks: towards the H-AID framework

The above review of various frameworks for humanitarian crisis analysis is summarized in Table 2.1. In Figure 2.5 we have translated this summary into locating each discussed framework in a three dimensional space outlining the three core qualities we used to evaluate these frameworks on.

From this, and the summary table (2.1), we can see what the various frameworks have to offer in relation to the ideal balance of core qualities which this chapter started out with. We see, for example, that both the PAR and SLF framework score quite well on offering programming advice and thoroughness, but they are quite slow to apply. PAR is not that generally applicable as the ideal would be, whereas the SLF framework is closer to this ideal. The political economy and arena approaches are much more generally applicable than the PAR framework and are closer to the ideal on the dimension 'quick and dirty' versus 'slow and thorough', but instead offer less concrete entry points for programming advice than for example the PAR approach. Whereas MIRA is generally applicable to various crisis contexts, it focuses on the immediate crisis phase and the phase preceding it, and is therefore quite 'quick and dirty', which also means that it has less to offer in terms of programming advice. The CVA approach is located a bit in the middle of all these other frameworks on all three dimensions.

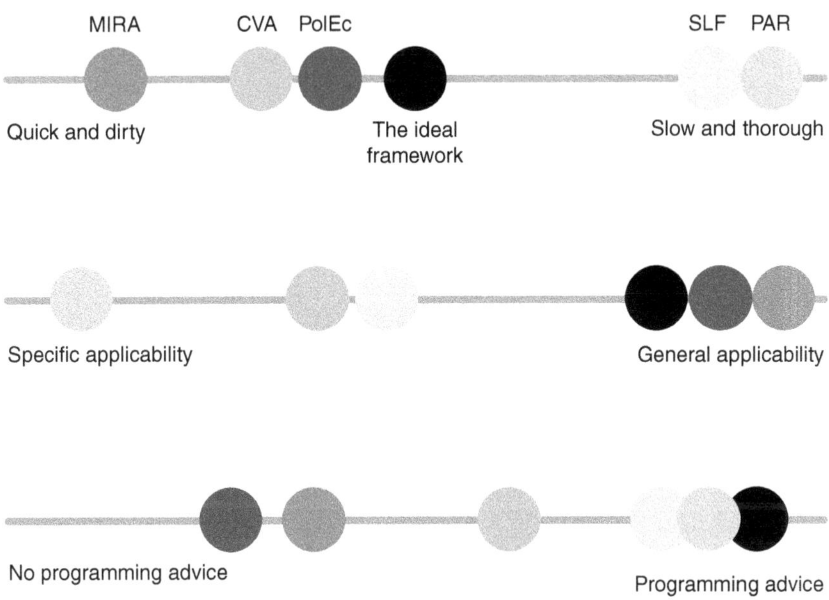

Figure 2.5 The five selected frameworks placed in the three dimensional space of core qualities.

Table 2.1 The five selected frameworks compared

Core qualities	*PAR*	*Political economy approaches*	*MIRA*	*CVA*	*Livelihood approaches*
Quick and dirty vs Slow and thorough	Slow and thorough	More slow and thorough than quick and dirty	More quick and dirty than slow and thorough	Quick and medium thorough	Slow and medium thorough
General vs specific applicability regarding various crisis phases and contexts	Quite specifically applicable: ‘only’ to disasters and in pre-disaster phases	Quite generally applicable in terms of crisis contexts, first only to conflicts, but now also applied to disaster settings. Especially useful prior and during humanitarian crises	General applicability to crisis contexts, but focuses on immediate crisis phase	Medium general applicability: mostly to disasters, but also as a prevention tool	General applicability in terms of crisis context. Mostly used as a more developmental approach
Programming advice?	The distinction between root causes, dynamic pressures, unsafe conditions and hazards, leads to potential for quite detailed programming advice	Provides some but not very detailed options for programming advice; especially useful for organizational risk and safety analysis and management	Helps to identify strategic priorities for programming, not meant as a tool to design tailored, localized responses	Quite basic programming advice, also reported problems with translating analysis to intervention	Possibilities for detailed programming advice due to degree of specificity of the model, however some criticism about feasibility

From the above it can be concluded that the selected frameworks present us with a trade-off between the thoroughness of the analysis and the time it costs to conduct a particular analysis, as expected. In addition it seems that the more generally applicable the framework – which would be a desired quality of any model – the less precise the programming advice becomes, which would be undesirable. With the H-AID framework we aim to present a middle ground by providing a framework that offers an acceptable degree of thoroughness whereas at the same time is it quicker to apply than the most thorough frameworks discussed in this chapter. In the same vain, the H-AID framework aims at being quite generally applicable (in terms of crisis phases and contexts) but also to provide tools for specific programming advice.

Next to the fact that the H-AID framework aims to offer a solution to the abovementioned trade-offs, it also aims to combine the most valuable structural dimensions from the frameworks discussed. As the above review shows, the selected models share a number of commonalities relevant for a structural approach to humanitarian action. First of all, most models focus on identifying the *capabilities and vulnerabilities* of groups as a core element of humanitarian crisis analysis, such as in the CVA, PAR, MIRA and SLF frameworks. Second, all models to some extent ask for an *analysis of crisis context dimensions*, something that is also recommended in the Good Practice Review on organizational security management (HPN/ODI 2010). Often these dimensions pertain to the social, economic and political fabric of the societies these crises occur in, complemented with physical/natural context characteristics. Whereas some models very extensively discuss these dimensions and their interconnections (the PAR and SLF model), other models provide more general elaborations of these dimensions and offer fewer options for in-depth causal analysis (MIRA, CVA), but all outline the importance of these context characteristics, also as a means to better understand the (direct) causes of crisis and vulnerability to crisis. Moreover, some frameworks recommend to conduct a pre-crisis and post-crisis analysis, such as the MIRA model. Third, most models point out the *importance of actors* (individuals, groups and organizations) in humanitarian crisis, although some refer to these more explicitly than others. Whereas the political economy and arena approaches put this at the center of their models, and explicitly include the analysis of power relations, other models include actors as one set of important dimensions to focus on (i.e. PAR, SLF and MIRA). Actors, their goals, resources and relations are deemed crucial to understand the causal dynamics of humanitarian crisis, but also to assess if and how the aid community and the work they do can become part of the power dynamics at play. This facilitates a risk assessment of potential adverse effects of aid (see also Anderson 1990) as well as organizational security analysis (HPN/ODI 2010).

The H-AID framework thus aims to build upon existing models by synthesizing and combining the core elements from existing models by elaborating three fundamental elements of safe and evidence-based programming: (1) context analysis prior and after a crisis hits; (2) making the step from context analysis to programming advice, including organizational security considerations; and (3) stakeholder analysis as a tool a humanitarian organization can use to identify strategies to achieve its intervention goals safely and effectively.

With regard to context analysis, the book builds upon existing elaborations of context dimensions (Chapters 6 to 10), by focusing on the economic, social, political, food, health and environmental (ecological) context and providing tips for indicators and data collection and analysis (see Chapter 3 for a more extensive account of how we came to these dimensions). In this analysis also a first identification of relevant actors in the various contexts can be of importance. In addition, it will be discussed how this analysis helps to identify capabilities and vulnerabilities of groups (Chapter 5). Especially with regard to the context analysis, we aim to present a middle range framework, generating sufficiently reliable and valid evidence to base programming decisions on, but also quickly enough to apply in emergency settings. We therefore also provide guidance as how to determine what information is valuable to collect, how to collect valid and reliable information, as well as to judge information quality, to analyze it, interpret it, scale and weigh it (Chapter 4).

In terms of connecting the context analysis to programming advice, a method will be presented how to translate the insights gained from the analysis into programming decisions by means of constructing program theories and investigating the assumed workings of an intervention prior to its start (Chapter 11). This is followed by the stakeholder analysis (Chapter 12), which we define as quite a specific type of actor analysis. In the context analysis one might already have identified a wide variety of actors involved in an emergency. With a stakeholder analysis we present a more focused method to analyze a stakeholder field regarding a particular issue from the point of view of one particular (focal) actor. In the case of humanitarian aid, this focal actor is often a humanitarian organization and the issue is related to the aim to intervene. In the stakeholder analysis we will elaborate how such a particular analysis helps to identify strategies such as how to achieve an aid organization's aims safely and effectively. In this analysis, one analyzes the goals and resources of, and relations between, a given set of actors crucial in achieving an aid organization's aims. Finally, as a crucial component of any intervention design, this book provides guidelines and theoretical considerations for monitoring and evaluation as well as quality assurance (Chapter 13).

In our view, the application of these elements (context, intervention design and stakeholders, next to monitoring and evaluation) as elaborated in this book will provide a sound basis for evidence-based programming and thus be helpful in generating impactful aid and a safe work environment for aid workers. In the following chapters, we will first elaborate the H-AID framework by outlining the principles of context analysis, its historical and theoretical background, and the various contexts to be analyzed.

3 Context analysis and securitization

Andrej Zwitter and Joost Herman

Often, context analysis is not seen as much more than simply providing a background on the history, politics and economics of a region. These elements definitely constitute key elements. However, a context analysis that is supposed to form the basis of decision-making in volatile humanitarian crises needs to be empirically and theoretically sound. This is why the NOHA Groningen team developed, among other tools, a comprehensive context analysis tool based on theories of security studies such as Securitization Theory and Human Security.

In this chapter, we introduce the Comprehensive Context Analysis (CCA) as a tool and explain the history and theory behind it. The theoretical foundation is based on the Copenhagen School of Security Studies as well as research on threat-perception and decision-making heuristics. Furthermore, we introduce the idea of levels of analysis in comprehensive context analysis.

Historical background

The traditional concept of security up until recently has always been defined in terms of states and the qualities of statehood. In such a traditional conceptualization states opposed each other, mistrusted each other and deemed it their prime task to survive by protecting their territorial integrity and the physical well-being of their citizens. As a consequence, security was, simply defined, the absence of physical threats to the territorial and functional integrity of a given state (Buzan 1991). Developments at the closing of the twentieth century, however, made it necessary to enhance this definition to do justice to a radically changed international environment.

The 'old-fashioned' almost exclusive emphasis on a state's means to survive in an anarchical environment and the importance of the inviolability of borders automatically dictated a build-up of forceful military machinery. Also, it was deemed of the highest importance that the population within the state was united and loyal to the state in its policy of defense. As a consequence, more often than not the well-being of the ordinary citizen was estimated to be of less importance than the means of protecting the territorial integrity. Even repression has often been justified out of reasons of real or imagined threats to the territorial integrity of states. In reality, political leaders used perceived threats to the country to consolidate their reign through repression of political opponents and their supporters (see for further information on territorial integrity, national unity and state power vis-à-vis citizens Steans and Pettiford 2001: 29–33, Cerny 2010: 48–50).

As of the 1980s, however, the security picture changed as a consequence of international developments, such as an expanding global market and the end of the Cold War (den Boer and de Wilde 2008: 9–17). An ongoing debate on security in a universal, regional, national and sub-national perspective has thrown aside the dominating interests of state security through military means. Olof Palme, former Prime Minister of Sweden and chairman of the Independent Commission on Disarmament and Security, was the first to emphasize the need to exchange the old-fashioned security concept for a concept of so-called 'Comprehensive Security' (Commission on Global Governance 1999: 44). In his commission's normative view sustainable peace and security could only be obtained if shared by everybody (human beings and their organizations, including states). Also, comprehensive security should hinge on the notions of cooperation based on the principles of equity, justice and reciprocity. As a follow-up the United Nations started to develop the concept of Human Security in which an exclusive emphasis is put on a human-centered approach of security, consisting of a secure environment for the individual, by which is meant, amongst others, food, social and health security (UNDP 1994). In short, human security refers to the quality of human life and respect for the human dignity. Only recently, both concepts were amalgamated, resulting in an analytic framework combining the human interpretation and evaluation of security with the 'classical' dimensions of state security: the 'Humanitarian Analysis and Intervention Design (H-AID)' Framework.

Development of a concept

The historical overview of the road to new concepts of security is a bumpy and diffuse one. At present, the ongoing debate is characterized by diverse ideas and conflicting sets of terminology, in which the protection against external aggression through military means still plays its part, but less so than it used to be. As a matter of fact, the mere importance of state boundaries has lessened to a considerable degree as a result of a growing transnational flow of capital and goods, increased means of rapid global communication, transnational coalition building and widening networks of interest organizations favoring the cause of norm convergence on human rights, rights of vulnerable groups and processes of international cooperation (Ohmae 1995, Martell 2010: 19–23). Also, within intergovernmental circles more and more the idea is accepted that the intergovernmental community is legitimized to interfere in the internal affairs of sovereign states whenever the quality of human life is threatened through (gross) human rights violations. Despite article 2 paragraph 7 of the United Nations Charter, emphasizing that the UN cannot interfere with domestic affairs, a (small) majority of states do not accept this article anymore for states to go about inhuman policies towards their own citizens without interference.[1] Kosovo (1999), East Timor (1999), Afghanistan (2001), Ivory Coast (2004), Libya (2011) and Mali (2013) are examples in kind. Nevertheless, many states vehemently oppose this growing practice of international interference. Finally, a third element adds to the complexity of the present day security debate, due to the recognition that the human component of security is growing ever more important, also the number of non-state actors in the field of security has augmented considerably.

The speed of the above-mentioned developments in the world in general, and in the security debate in particular, have, by and large, been influenced by five major political events in the 1990s and one in the twenty-first century.

First of all, the end of the Cold War lessened the willingness of the United States and the Soviet Union/Russia to meddle in the conflicts amongst their former clientele. As a consequence, the number of conflicts (and human tragedy) has risen considerably in Africa, Asia and south-eastern Europe, whereas in the days of the Cold War they would have been contained. Especially in the case of so-called 'low priority conflicts' the United States, Russia and their major allies seemed to use an adapted version of Mao's adagio 'Let thousands flowers blossom': 'Let thousands conflicts blossom'. Only recently, one can discern great powers taking once more an interfering stance towards conflicts in their immediate vicinity. Especially China and Russia and their respective 'back yards' can be mentioned (see for further information on blossoming conflicts Dijkzeul and Herman 2011, and on reorientation in foreign policy Medvedev 2004: 55–7, Jisi 2011).

Second, the so-called third wave of democratization in the former socialist world and the Third World has caused much confusion and many feelings of insecurity (Diamond and Linze 1989, Volten 1992). In some countries the growing number of actors at the national and sub-national level – situated in a pluralist framework as a result of political changes – undermined national consensus on topics like national security, let alone on international security models. The 2014 split position within Ukrainian society which way to turn – West or East – sums it up and even has brought the country to the brink of implosion (de Deugd 2005: 97–101, Roslycki 2011: 123–30). In other countries – such as Somalia – sheer chaos rules as a result of which security at the national level actually decreased as opposed to the previous situation in which authoritarian or semi-authoritarian politicians ruled these countries. The most tragic and ultimate consequence hereof is the so-called failed state, where central government just does not function anymore. The danger of transnational contamination in these areas is very real indeed.

Third, and closely intertwined with the previous arguments, the role of the state as sole actor in the field of security has diminished. Other actors have started to play their role, like multinational corporations, ethnic groups whose members live dispersed over more than one state territory, nonprofit non-governmental organizations, religious sects and even terrorists, to name a few. The activities of these groups, although in the majority of cases of a benevolent nature, are hard to control also if developing in a malicious direction (the greed for profits, intentional pollution of the environment, intolerance towards minorities, outright xenophobia, all negatively influencing the feeling of security) (den Boer and de Wilde 2008: 9–11).

Fourth, the development of the concept of interdependence between states and transnational actors has diminished the value of state sovereignty and therefore the dominant role of the state in the security debate. Modern means of communication, capital flow, ideas on the universality of human rights and political and ecological vulnerability in general are the causes (Brown 2007: 171–3).

Finally, the 2001 terrorist attack on the twin towers of the World Trade Center and the Pentagon by the use of aircrafts has caused a renaissance of a state-centered approach of military security and state security in the area of policing. The effect of this renaissance can be observed at the acts of parties to the Global War on Terrorism (McGrew 2007: 24–5).

As a result of the combined developments outlined above several threats related to humanitarian aid provision can be listed:

- intergovernmental and non-governmental interference in the domestic affairs of states, which more often than not causes friction between states and worsening of internal repression;
- proliferation of conventional armaments, for example land mines and chemical/biological agents through former socialist states, Russia, the United States, France and Great Britain, the effects of which have been visible in a variety of African states but also in Syria;
- gross human rights violations at the national level, especially in failed states and in those countries where leading political elites successfully oppose democratization and crush popular risings (Somalia, Yugoslavia, Bahrain, Syria);
- internal power struggle resulting in a permanent repression of the population;
- economic deprivation and denial of the individual right to own property;
- threats to social security in general (in the field of security of existence, integrity of social networks, 'gender issues' and health care security);
- threats to the human ecological habitat (undermining people's health, the right to property and food security). This category can be subdivided into the categories complex natural disasters (e.g. volcanic activity, earthquakes), complex political disasters (e.g. starvation of a population as a consequence of deliberately destroying crops by warring factions) and a combination of both (flooding as a result of irresponsible use of environmental resources).

In conclusion, the debate on security has been broadened, next to the classic element of territorial integrity, to include the physical and mental well-being of populations, a process strongly influenced by transnational and even universal developments. In the words of the UN-Commission on Global Governance, urging for a new and universal concept of security given the changed universal environment, 'lasting security will not be achieved until it can be shared by all, and that it can only be achieved through co-operation, based on the principles of equity, justice and reciprocity' (Commission on Global Governance 1995: 46). As a consequence of the changed nature of security and the enlargement of security criteria, also the number of actors involved has increased considerably and their *modus operandi* has changed as well, influencing the mode of analysis of H-AID.

Context analysis in the H-AID framework: dimensions and levels of analysis

The H-AID approach is related to the developments in the security debate as discussed above as it applies similar context dimensions. The recent trends in the coordination of humanitarian action acknowledges that in any man-made, natural or complex emergency a multi-sectoral approach (on the level of OCHA this is the cluster approach) needs to be employed in order to comprehensively treat the root causes of the problem and not only a few isolated symptoms. This cluster approach reacted also to the proliferation of humanitarian actors that had led to overlaps in some areas (for example many actors distributing food) and crucial gaps in others (lack of aid in the area of water and sanitation). Moreover, it is believed that such an approach more sustainably addresses the emergency, enhances accountability of the actors involved and increases predictability of the outcomes (OCHA 1999, 2006). The H-AID approach was developed, first, to allow for such an overall assessment. At the same time researchers commonly raised the criticism that a wrong modus operandi in giving aid could actually worsen the situation of beneficiaries instead of improving it (de Waal 1997, Anderson 1999). Thus, second, our approach also aims to provide the right tools to recognize those social mechanisms that could put the provision of aid in jeopardy and those that can be utilized to improve the efficiency of aid. Both goals in combination lead to an overall analytical framework that may contribute greatly to organizational learning. Two components are central in achieving a comprehensive context analysis, i.e. the identification of context dimensions and of levels of analysis.

Context dimensions

The modern concept of security in analogy to human security goes beyond a state centric and military approach to security. Instead, it 'securitizes', i.e. treats as security threats, different agenda points of the international community, such as environmental, economic or health issues, by emphasizing the need to secure these issues for individuals in order to enable them to achieve a reasonable livelihood. According to the United Nations Development Programme (UNDP) *Human Development Report* (1994: 23) – one of the first documents in which human security was presented as a core concept – it 'means, first, safety from chronic threats such as hunger, disease and repression. And second, it means protection from sudden and hurtful disruptions in the patterns of daily life – whether in jobs, in homes or in communities'. This protection of the individual is reflected in seven context dimensions as defined by this report (UNDP 1994):

- economic security
- food security
- health security
- environmental security
- social and cultural security
- political security
- personal security

A common criticism is that these dimensions of security as defined by the UNDP are overlapping or even redundant as, for example, food security is mostly a question of economic security.

Levels of analysis

Another way of looking at security is the Copenhagen School that divides security into different sectors, which in turn can be analyzed on each level from the international to the regional to the sub-state and even to the individual level by dividing it into the sectors (Buzan *et al.* 1998): military security, environmental security, economic security, societal security and political security. Inherent to all these categorizations of security is that by categorizing it, security is automatically fragmentized and critics would see in unavoidable overlaps a weakness to the concept as such.

Context analysis according to the H-AID framework follows the views of these two approaches to a certain extent as it builds on the different dimensions of security as basically elaborated by the UNDP and on the idea of allowing the different levels of assessment as pointed out by the Copenhagen School. We follow this line of reasoning and therefore distinguish between the following levels of assessment:

- the international level (international community and organizations);
- the regional level (regional organizations);
- the national level (states);
- the group level (organizations, communities, social groups);
- the individual level.

For the assessment of a context with the H-AID framework in humanitarian action NGOs would for example mainly apply the analysis on group, individual and national level when planning projects, while international organizations would mainly refer to the national level and higher levels when developing programs.

The above-mentioned levels do not reflect a normative view on what the defining level of analysis *should* be, or which level is most important; it reflects the potential levels of context analysis in the H-AID framework. Depending on the specific issue of attention, as well as on the disciplinary background of the researcher, a particular level of analysis can be deemed more important than another, or can serve as a first point of departure for analysis. For example, political scientists will be inclined to focus on the national level of analysis, international relations scholars on the international level, whereas sociologists and anthropologists are more inclined to take groups and individuals as their point of departure for analysis.

Overlap in context dimensions and levels

As stated previously, the above divisions into dimensions and levels might leave the impression of overlap. How to avoid the problem of overlapping context dimensions? The solution is simply the answer to the question of why the different dimensions are overlapping. The term 'security' describes a condition that enables actors to proceed in their actions as perceived as 'normal conduct of actions'. If a normal conduct of action is not possible, then security, as a general conception, is restrained by certain factors. These factors are thus described as security threats. To give an example: democratic states see free elections as the condition that allows the functioning of a democratic state; if free elections are not possible due to a state of emergency, one would speak of a threat to political security. In essence, when people use the term security they refer to the condition of an optimal (or at least normal) conduct of actions without major restrictions.

The different dimensions of security are nothing more or less than the description of this condition in a specific field of interest (as above political security). These fields of interest are, however, purely arbitrary; meaning, every noun that describes a category of things can be meaningfully combined with the term security. Since the contextual dimensions are all interconnected – as they are always relevant for the chosen level of assessment and the chosen category of security – it will not surprise that they overlap. Human security is about the condition of optimal conduct of action of the category 'human'. State security is about optimal conduct of actions of the category 'state'. The H-AID framework unifies both concepts as it describes them as mutually conditional: the actions of a human are usually embedded in his environment of social ties and ultimately (but not always) the state as a construct of social ties with the element human without which a state is unthinkable.

So when human security is described in various dimensions (as: economic security, food security, health security, environmental security, physical security, community security and political security) it is still about the category 'human'. The category 'human' can, in its needs for keeping up the state of a normal conduct of actions, be summarized as purely material needs for keeping up body functions as well as needs that depend on the social environment like social inclusion, intellectual or cultural exchange, political participation etc.[2] Many of the material needs also have to be acquired via the social environment. The human as social entity is therefore always dependent on his social environment that defines the human's security situation. For example, if the social network of a society has collapsed due to civil strife, this might result in a limited ability of the community to provide services and items necessary to secure the food and health contextual dimensions.

In the Western world, the social environment from where a human derives some of his basic needs is often in functioning states. It is these state structures that also contribute to the human security of individuals. Hence, if the functioning of the state is threatened in dimensions that also describe human security, then – due to mutual dependency – a threat to state security is at the same time a threat to human security. In such contexts, these overlaps of dimensions of human security and state security are only natural because the dimensions are largely empirical and describe necessary standard conditions of human survival following from this state survival and vice versa. By analogy of this mutual dependency, the extension of state/human security to the various levels is a logical step. That means that the protection of all contextual dimensions on each level is vital for achieving a general (and thereby ideal) level of security in all context dimensions. Following this logic of interdependency on all levels of security, it becomes visible, that if for instance the state as a security provider for its population is missing, people on the group and individual level have to ensure their survival only within the same level – the state as a link between group and individual level and regional and international level is missing in these cases. Only occasionally, security for the people on group and individual level is provided from actors at the regional or international level as in the case of international humanitarian assistance or humanitarian intervention. This kind of action deployed by other actors above national level (by IGOs) depends on whether the state is unable or unwilling to ensure a comprehensive security of its people. It also depends on the extent of the threat, media coverage and political interest, amongst others, how well the international community is informed about ongoing crises.

For the purpose of analysis of an ideal state as defined by the comprehensive context dimensions the following assumption applies: if there is no dimension of human survival and state survival missing in the list of dimensions that are covered by the H-AID framework, then these dimensions of security are useful for an analysis of the state of the contextual dimensions on all levels of assessment. Hence, state security as well as human security might be a matter of concern and a subject of analysis on all levels, as the following scheme (Figure 3.1) visualizes. It should be noted that in social reality there is not such a natural and ideal relationship between state and human security. For example, there is ample evidence that states can hamper or damage the security of groups or individuals within their borders. The distinction in various levels of analysis exactly points our attention to this: it shows at which levels the preconditions for a secure context are absent or present.

Looking through our prism, we can now assess every contextual dimension through the light of one dimension. That means, we can look for example at health security through the light of economic security and might see that the health problem encountered in the case study is not only a health problem per se but also an economic problem and the economic problem could be again a

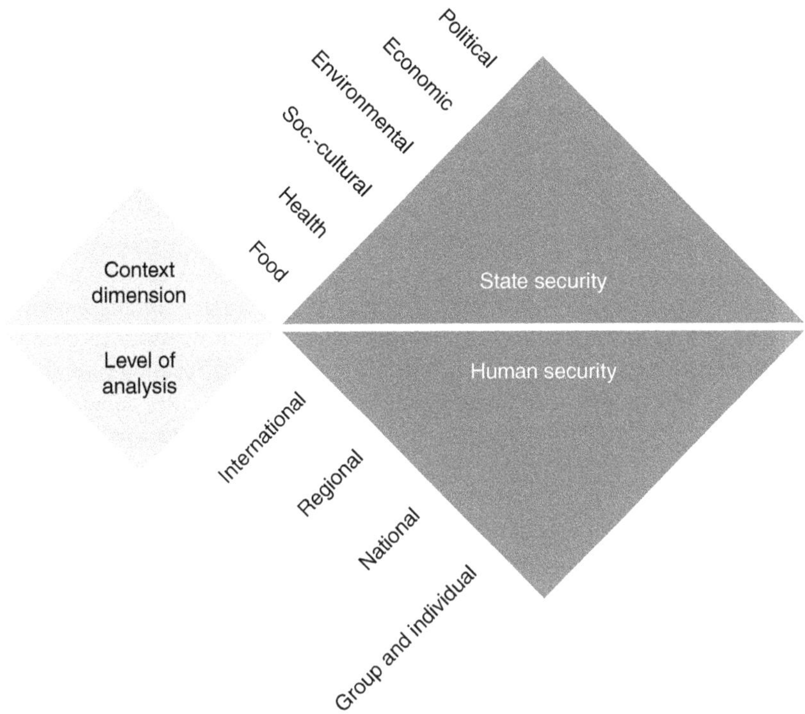

Figure 3.1 Levels of analysis in H-AID.

societal problem. Hence, by taking a look at the dimensions of security through the light of different context dimensions and thereby through the glasses of different disciplines of humanitarian studies, one can see each dimension of the contextual complex of the H-AID analysis as a factor of security of the category 'human' as such. This leads to the assessment of each of the six dimensions through the light of the other six dimensions. This enables (or better, it demands) a multidisciplinary assessment of security.

Understanding contexts through securitization[3]

In the H-AID framework we propose to use a so-called objective scaling technique of *absolute scaling*, as will be elaborated in Chapter 5. It is, however, important to understand that behind what appears to be an absolute and objective measurement are relative perceptions of different actors. The theoretical mechanism at work behind the apparent (read: perceived) objective context of humanitarian crises is 'securitization'. Crisis contexts are typically perceived as an environment, which encompasses threats to certain aspects of security (a referent object). These perceived threats to a referent object (e.g. environment or health) are used to justify extraordinary measures through speech acts, i.e. an utterance with performative function (Buzan *et al.* 1998, see also Austin 1965 on speech act theory).

Due to the perceptive nature of threats not every context relevant to the individual and the smaller groups can easily be objectively measured with quantitative and qualitative data. One has to deal with the cultural settings that define insecurity in order to assess a situation. That leads to a complicated problem: threats perceived by the analyst are not necessarily likewise perceived by a local community. This might be because their cultural setting reflects different values or because the knowledge necessary to perceive something as a threat is missing. For example, the lack of knowledge of the threat of a volcano eruption in a referent group, would mean a lack of securitization, whereas a researcher equipped with this knowledge would try to securitize it. Likewise, if local culture argues that the volcano eruption is an act of godly punishment, the securitization responses triggered might be different in the local community (prayer) than that of the analyst (evacuation).

Another factor that influences the context analysis is that by researching such situations the analyst changes its environment by pointing to what she perceives as threat. By interacting with the reference group for her inquiry the analyst creates awareness among the researched group of a possible threat. This awareness in turn is being interpreted by the reference group through their cultural lens and either will be accepted, distorted or disregarded.

Adding to the problems with the perceptive nature of existential threats is the question of establishing thresholds between basic needs on the one end and luxuries on the other, and between acceptable risk and intolerable deprivation. It seems necessary to add thresholds to the security discourse assessment that can be measured objectively, like numbers of deaths related to perceived threats. One might assume that insecurity can be objectively measured by the amount of deceased persons based on the key assumption that the more people die the higher the likelihood of a situation to be prioritized as an existential threat. A simple example, however, proves this key assumption wrong. In 2005, 41,600 people died in traffic accidents in the EU, while in the same year 56 people died in Western Europe from terrorist attacks (Coolsaet and van de Voorde 2006).[4] While a major political discourse is going on regarding terrorism, the economic crisis and other pressing issues, road safety is hardly securitized (although a policy issue).

Numbers of casualties do not always represent what is perceived as threats that justify extraordinary measures. It seems in this case that either we accept the road safety situation as unchangeable, self-inflicted or not threatening to the political system or justify it simply as a necessary risk to be taken given our dependency on cars as means of mobility (see Mueller 1990). Another explanation can be that threats related to 'abnormalcy' are securitized more than threats related to 'normalcy', e.g. death by car accident belongs to the daily experience in Europe, death by terrorist attack does not. As a consequence, the societal and political impact of terrorist attacks diminishes as their frequency grows. That the

Israelis have come to live with the omnipresent threat of suicide bombers is a telling example. Adjustment to street violence in most megalopolises of the world illustrates this, too. Still another explanation can be that potential securitizing actors have nothing or little to offer to counter the threat of traffic death (it requires costly readjustment of the entire society), whereas securitizing terrorism opens profitable new windows of opportunity for different branches of the war economy.

The above discourse-based approach focuses on speech-acts, it does not ask about the objective threats, but what securitizing actors put on the agenda. One would think that when objectively basic needs are not met, this will be perceived as a threat and therefore be reflected in the local security discourse. However, it happens that these issues are not mentioned at all. Take for example domestic violence: the mass of women in patriarchal societies do not tend to express what the analyst will see as a security problem, and what in international society is defined as violations of human rights: rape, repression and forced labor. Possible explanations are: the women do not want to make themselves a target by bringing up violence as a topic of political discussion, or they are kind of hospitalized in their situation, they do not see an alternative, or they simply consider their situation normal. Habituation, fear and uncertainty about alternative paths may outweigh the certainty of survival in miserable circumstances.

The status of insecurity assessed according to 'objective' indicators (as set, for example, by the UNDP) runs the risk of representing mainly the values relevant for the analyst. This assessment is not necessarily relevant for the reference group, if they would be asked to comment on that. It means that we have to make use of indicators that allow the assessment of taboos, which are reflected in actions rather than in words, and of the unidentified problems that will have an impact on the group nevertheless, like, for example, health. The spread of deadly disease, even though not perceived as a threat because there is no knowledge about it (HIV-Aids in Africa particularly before the worldwide campaign to reduce it started there), is still a threat to security as defined by the analyst. By raising this threat as a problem in the context analysis, the analyst then becomes a securitizing actor herself, something that one should be aware about.

Threat perception and securitization

What individuals in their social environment identify as risk or threat is learned by direct or indirect experience linked to expectation. This can take the form of *comparison and prognosis* or of *abstraction and prognosis*.

Abstraction and prognosis

The way abstraction and prognosis works, is well explained with the example of the nuclear holocaust. During the Cold War the securitized threat was a nuclear holocaust. We never experienced it, but imagined it (helped by Hollywood, peace research and Cold War game theoretical modeling of scenarios). We were able to take what we learned and communicated about the effects of nuclear warfare, abstracted it and imagined what would happen, if it was waged – the power of imagination. Another issue-area where abstraction and prognosis figure strongly in securitization processes is in the environmental context (Buzan *et al.* 1998, de Wilde 2008): climate change, its origins, nature and consequences, is about the credibility of models that predict future environmental catastrophes.

Comparison and prognosis

Other types of environmental threats, such as natural hazards, industrial disasters, desertification, deforestation, depletion of resources and species, tend to be in the realm of comparison and prognosis. This is the case when group A compares oneself with another group B, or with oneself at earlier points in time and projects the observed trend into the future. If this projected future would then hold a believable hypothetical threat it can serve for securitization purposes. For example, oil is still available but states securitize it with a view to the depletion of the reserves in the future.

One key assumption that guides the analysis of the context and the reaction of actors to threats is based on theories that attest basic needs as the impetus for human behavior (Maslow 1943, Lasswell and Kaplan 1950). It theorizes that when people are threatened in their basic needs or values, this dominates their behavior. Any acute threat to survival would fall into this category. *A maiore ad minor* it means that the probability of the emergence of a security discourse that justifies extraordinary measures regarding a specific aspect is the more likely the more basic needs are affected. This key assumption acts as the 'objectively' set threshold applied by the analyst. A method that allows the measurement of basic needs enshrined in human security has been developed by Owen (2002). He advocates measuring human security on the basis of a body count related to different contexts. Though observing our limits, a body count can serve to support this key assumption. Acute threats to survival seem to be a universally applicable threshold where a securitization process is to be expected. We attest this threshold universal applicability.

Above the threshold of survival and basic needs, empirical measurements will have to resort to the concept of *relative deprivation* between a reference group and a comparison group with relevancy to the reference group. The concept of relative deprivation is used to explain the tendency for political violence conducted by an actor depending on how extreme this disparity of capabilities and expectations in the area of her basic needs is perceived (Gurr 1974, Stewart 2008). This comparison group can either be a group that is in geographic relation to the reference group ('why can't we have what they have?'); or it can be a historical comparison group ('we did better in the past, let's restore it'). This means that the reference group compares itself with itself at an earlier point in history or with another closely related group experiencing better circumstances.[5] The higher the disparity between the situations of comparison, the higher the likelihood of a security discourse, even if there is no or little evidence of the factors described in this key assumption. This approach allows us to expect a securitization process on the basis of data, like income or crime rates or living standards in general.

Effects of abstraction and prognosis versus comparison and prognosis

A factor involved in abstraction and prognosis might be the 'Peak-End rule' (Fredrickson and Kahneman 1993). It explains why a slow increase of casualties due to the accelerating effect of climate change on natural disasters (comparison and prognosis) tends to lead to a less steep securitization process than abstraction and prognosis. Peak experiences, such as 9/11 or Hiroshima and Nagasaki, are felt more strongly than slow changes (e.g. climate change). This seems to be so, even if the final outcome of slow changes might be the same or worse than the result of peak experiences. Abstraction and prognosis can then lead to an overweighting of the probability of events with high impact (in the case of terrorism the high-impact/low-probability event of 9/11 led to massive securitization).

Another study discovered a cognitive error of judgments: humans tend to neglect situations with a continuous duration of pain (Kahneman *et al.* 1993). 'Duration neglect', as it is called, might explain why car accidents, which happen on a daily basis, are not as strongly securitized as terrorism occurring only incidentally in Europe and the United States. Comparison and prognosis, therefore, has a less strong effect on securitization similarly (see Taleb 2008: 76–9).

The mechanism of 'duration neglect' leads to no or unsuccessful securitization. The 'Peak-End rule' distinguishes between low securitization as a result of comparison and prognosis and high securitization in cases of abstraction and prognosis.

Conclusion

Context analysis is vital for any project if it wants to produce predictable results. To simply look at a situation and write down what one sees is, however, not sufficient at all. This became clear when demonstrating the utility of separating the context into different sectors.

Although the framework for context analysis in this book proposes to use an absolute scaling technique, which may leave the impression of objectively assessing a crisis context, we ask for caution because the discourse about what are the indicators for different contexts is a question of letting the analyst decide, by academic method or discourse, on the broadest possible description of an aspect, and letting the reference group decide, by expression of perception through their words and actions, which indicators have to be used. So the analyst will expect certain aspects to be important beforehand. Security then is about how actors relate to each other and to their natural environment in terms of threats and vulnerabilities.

While absolute scaling seems to suggest that within a context threats to certain sectors are objective, the analyst needs to remain aware that there is always a difference between how she perceives a context and how a local community perceives it. Furthermore, the analyst is now able to differentiate the effect of abstraction and prognosis on actors (stronger inclination to securitization) from the effect of comparison and prognosis (weaker inclination to securitization). The kind of securitization that an actor is affected by can be very informative for the stakeholder analysis (see Chapter 12), particularly when estimating the salience of actors vis-à-vis achieving a certain goal (e.g. abstraction and prognosis tends to lead to a much higher salience to securitize and act against a certain threat, comparison and prognosis). Before we now outline how to exactly compile information collected to conduct a context analysis based in the proposed dimensions and levels, we first discuss issues of information collection, reliability and validity – all crucial to a valuable and meaningful context analysis.

Notes

1 Article 2 para. 7 reads:

> Nothing contained in the present Charter shall authorise the United Nations to intervene in matters which are essentially within the domestic jurisdiction of any state or shall require the Members to submit such matters to settlement under the present Charter; but this principle shall not prejudice the application of enforcement measures under Chapter VII.

2 Gurr (1974) asserts that values, which constitute the goal objectives of human motivation, are presumably attributable to or derive from basic needs or instincts. Here he refers to Henry Murray (1938), who listed 12 viscerogenic and 28 psychogenic needs, which he categorized into three groups as: welfare values, power values and interpersonal values.

3 Parts of this chapter are based on collaborative work with Prof. Dr Jaap de Wilde.

4 See also 'Road traffic deaths and injuries', www.erso.eu/knowledge/content/09_post-impact/road_traffic_deaths_and_injuries.htm, accessed April 18, 2014.

5 This follows Weber's definition of social action that can be geared towards the past, present or future in contrast to reactive action that is geared towards the present (1978: 22–4).

4 From theory to analysis

H-AID methodology

Rafael Wittek and Andrej Zwitter

This chapter discusses some practical issues related to the collection of data in humanitarian contexts, and the assessment of its reliability and validity. Since space limitations prohibit an extensive treatment of these methodological issues in this book, this chapter can only provide a fairly brief and superficial sketch of some of the most important issues. The interested reader may consult the relevant literature for more detailed introductions. There are many excellent textbooks and guides on these topics, e.g. Bernard (2011).

The next section sketches the general purpose of our evidence-based H-AID framework, i.e. to move from an abstract and general understanding of the omnibus context to more domain or security specific theoretical constructs that can be subjected to empirical proof. The second section of the chapter describes the steps that are necessary to translate the theoretical context dimensions into valid indicators. The third section addresses the question how to identify trustworthy sources for these indicators. Finally, the fourth section discusses the question how to arrive at accurate estimates from trustworthy sources.

From omnibus context to theoretical constructs

In order to deal with information more effectively, it is good to get a general idea about context analysis. The most general form of context analysis is the omnibus context – in intelligence analysis this technique is also called starbursting (Heuer and Pherson 2010: 102) referring to the following questions: Who? What? When? Where? (To Where? From Where?) How? Why? This technique requires sorting available information in accordance with generic criteria that are present in all contexts. Further questions more specific to humanitarian crises would be: What do we know about the causes of the crisis? What do we know about the consequences of the crisis? What do we know about relevant actors?

In addition, two other important questions to ask are: What are the known unknowns? What are the unknown unknowns? These two questions deal very specifically with present and lacking information. One can distinguish between 'known unknowns' (I know that I need to know but I have no information) and 'unknown unknowns' (I don't know that I need to know and whether I do or don't have that information). The known unknowns are quite easily responded to. If one knows that one is lacking information about malnutrition, one knows what data to collect. The unknown unknowns are a little more troublesome. Sometimes crucial information that could decide over the success or failure of a project is missing simply because of a general unawareness that a certain issue is relevant in the given context. This usually means that one knows too little about a given context in order to reasonably predict the effect of an intervention. For example, one might be well aware about all nutritional facts to implement an emergency feeding project, but one might not know enough about the cultural context that might prohibit certain feeding practices or not enough about the security situation in the region to conduct the project safely. The cultural element of feeding practices will generally be a known factor to nutrition experts; that is, it will more likely be a known unknown. However, the relevance of the security element might, due to the focus of expertise, completely escape the expert (unknown unknown).

In order to reduce these unknown unknowns, one should generally conduct a thorough analysis of the omnibus context to pick up hints of potentially relevant unknown factors. The tools presented in this book may help to realize this objective.

From theoretical constructs to valid indicators

Political, economic, social, food, health, environmental security … the starting point for a context analysis is not only *abstract* theoretical constructs, they also represent *compound* latent dimensions, i.e. they usually contain more than one facet or sub-dimension. This means that in order to make verifiable statements about the level of security threat in a domain, four operations are necessary.

Decomposition

First, one has to decide which sub-dimensions are best suited to map the general construct, i.e. one has to decompose the overarching construct into more specific categories (decomposition). For example, two possible sub-dimensions of the theoretical construct 'food security' could be to what degree (1) everyone in the population has sufficient access to food, and (2) the quality of the food is sufficient. Ideally, the number of sub-dimensions covers those sub-dimensions that are necessary while at the same time being sufficient to adequately encompass the overarching construct. Determining when the sufficiency criterion is met, remains a subjective exercise for most security domains that do not dispose of inventories of pre-defined diagnostics that have been agreed upon by a profession (e.g. like is the case for medical doctors). This judgment will also be affected by effectiveness considerations. Effective assessment requires that one should collect neither too much, nor too little information. Too much information will be hard to quickly assess and sift through. Too little information will be missing the relevant data and make the conclusions susceptible to unavoidable biases in the base data. In other words, one needs enough data to check on uncertain sources. The sources as well as the validity of the data (i.e. how well the data describes what the analyst wants to know) also affect how much (additional) data one needs. The lower the reliability and validity scores regarding particular resources, the less confidence the analyst can have in the assessment.

Operationalization

Second, one needs to find out what would be valid indicators to measure each sub-dimension (validity), and how many indicators would be needed to sufficiently cover all relevant aspects of this dimension (reliability)?

Information and evidence can be called *valid* if it is relevant and represents what one wants to know. Some indicators may simply not be valid measures for a construct. For example, if one wants to know how many people in a village are in need of what kind of nutrition, knowing the average nutritional levels of a region does not tell much. In other cases, a set of indicators may be valid, but some of them might be better and others worse in describing a certain context dimension. For example, both the average caloric intake of a population and the vicinity of food distribution points may be indicators for the sub-dimension 'access to food'. But the vicinity of food distribution points does not say much about the availability of food or the target population's ability to acquire it. Consequently, this indicator has validity problems, particularly when compared to the more direct measure 'average caloric intake'. However, this does not mean that vicinity to food distribution has to be excluded as an indicator: depending on the specific crisis situation, this indirect indicator may therefore still cover a crucial part of the 'access to food' dimension, and could be incorporated (eventually with a somewhat lower 'weight' than the more direct indicators, see below).

Operationalization of more than one indicator to measure a sub-dimension is a crucial step to ensure the *reliability* of an estimate. To take the analogue of a medical check-up: the physicians' verdict about the health condition of a patient usually is based on more than one kind of measurement (e.g. heart rate, lung function, a variety of different blood values). The reliability of the physician's statement about the health condition of a patient increases with the number of valid indicators that enter the overall assessment. Similarly, if one decides that food access is an important dimension to analyze needs of people struck by drought, then there might be different ways to measure this – e.g. one could measure household income and food prices on the market.

Finally, a note on predefined lists of indicators. One should be aware that such lists of indicators, though useful as a general guideline as argued in the following chapters, also have some major limitations. Indicator lists are never complete and not applicable to all times and places; this depends on the situation that is assessed, as well as on the availability and reliability of information.

Quantification

Third, in a real life exercise, most likely the information available for each indicator will be highly diverse, varying from census data collected by national bureaus of statistics, to incidents reported in the local newspaper. Sometimes, all that may be available are subjective reports from informants. Put differently: what enters a context analysis usually is data representing at least three different measurement scales: (a) interval scales (e.g. expenditures in dollars); (b) ordinal scales (e.g. low, medium, high); (c) nominal scales (e.g. presence vs absence of something). During the quantification step, the analyst needs to translate this heterogeneously scaled data into the measurement scale of the specific context analytical tool to be used. For the Comprehensive Context Analysis (CCA), this is an ordinal scale ranging from '0' to '6' (see Chapter 5), distinguishing conditions of security from conditions with latent or manifest threat. For the stakeholder analysis, these are interval scales ranging from '0' to '100', reflecting e.g. the degree of influence of a stakeholder (see Chapter 12). It is evident that this operation – in which the analyst has to map a qualitative observation or a quantitative fact into a number – entails a lot of subjective judgment, and is of course vulnerable to measurement error. Yet it is this step that forces the analyst to explicate the underlying assumptions, and to transparently document them. Therefore, when working with this quantified information, one should always keep in mind that each number is not more than a placeholder for a qualitative and highly subjective judgment. The main purpose of this quantification step is to facilitate systematic context analysis through comparison both within and across context dimensions, time periods (i.e. pre- and post-emergency) and different crises situations.

Aggregation and weighting

Fourth, it has to be decided how these indicators are related and 'add up' to provide an overall estimate for the overarching construct, i.e. the final assessment for the level of security threat in a specific domain (aggregation). For this step to be carried out, it is necessary that all previous steps have been completed, i.e. that each indicator of a sub-dimension has been translated into a value on the respective (security or stakeholder) scale. Once this has been done, the different values can be aggregated. Though adding them up and dividing by the number of indicators is the most straightforward and most often used technique of aggregation, more refined approaches are sometimes even necessary. The reason is that in the case of simple addition, equal importance is attributed to every single indicator. Whereas this may be warranted in some cases, it may heavily distort the picture in others. In the latter case, introducing weights may be an option.

Weighting is an instrument to adjust the relative importance of an indicator in a set of other indicators measuring an overarching construct. For example, when assessing the level of political security in a region, an analyst might consider the following indicators: level of freedom of the press as measured through incidents restricting freedom of speech; level of human rights as measured by the number of documented tortures; level of trust in institutions as measured through survey data. The analyst may consider all three dimensions as equally relevant and as valid sub-dimensions and therefore may decide to add up the values of the three. But the analyst may also attach different degrees of relative importance to each of them. For example, he or she may consider the presence of human rights violations as more important than infractions of the freedom of speech and institutional trust. In this case, the analyst may decide to weigh the human rights indicator twice as heavy as the other two, e.g. by multiplying this indicator by 50 percent and each of the other two by 25 percent.

From valid indicators to trustworthy sources

Once it has been decided which indicators are valid, one can look for potential data sources that may generate the desired empirical evidence. Trustworthy sources and reliable data are a major precondition for a good context analysis. A source is 'trustworthy' if the information it provides is not biased towards particularistic interests or is deliberately misleading. The information a source provides is 'reliable' if it accurately describes the respective facts. Note that a source can be trustworthy but at times deliver unreliable information, for example because it may only have partial information on a particular issue. Similarly, there may be cases in which sources that are considered as not trustworthy may provide reliable data, for example because it may be in their interest to do so in specific situations. This implies that the analyst needs to carry out independent checks, one on the general trustworthiness of the source (e.g. based on the source's reputation), and one on the reliability of the information provided by a source. We address trustworthiness issues in this section, and data reliability issues in the next section.

There are a variety of different data sources, ranging from informants, written media (e.g. government reports, official statistics, newspapers), to the various sources on the internet (e.g. blogs, interviews, wikis). Faulty or misleading information can jeopardize the success of a project and the reputation of an organization. Sometimes, deception is exactly the aim of the source, sometimes wrong information is simply the result of human error and at other times it is the result of a bias in the source. How to use the available data in such a way that a reliable and valid analysis can be made? Here we present some first steps that facilitate the correct selection of information to be used for assessing levels and degrees of comprehensive security.

After having made an inventory of potential information sources, one has to do a quick assessment about their trustworthiness. The following questions could help the analyst in this regard:

- Is the source generally trustworthy? For example, are the authors transparent in how they collected the information (method), what their sources were, who were their respondents?

- Is the source independent from partial stakeholders?
- Is the data and/or intelligence presented coherently with other assessments?
- Is the way the data is presented emotionalized or in any other ways biased?

The aim of this assessment is to categorize your sources from most trustworthy to least trustworthy. For example, data coming from an authoritarian government denying the outbreak of cholera in its own country (maybe to avoid an influx of foreign agencies) is of much less value than data presented by an IGO or independent academic institute that claims that there is an outbreak of cholera. Categorizing your sources in this way helps you to deal with such inconsistent or conflicting information. In the ideal situation one only works with the most trustworthy sources; however, if certain information is needed, then one could work with less reliable sources. However, one then needs to carefully assess the potential bias in the information presented and to account for this bias in the analysis. The following verification techniques can help the analyst to classify the reliability of sources:

- background check of the data provider;
- cross-checking the information with the same data of a different source;
- cross-checking the information with different data of a different source.

Tables 4.1 and 4.2 give a more detailed overview of key elements to check when assessing the reliability of sources and the accuracy of data.

The criteria specified in the tables can be used to provide an overall assessment of data quality. 'A1' ratings would reflect the highest quality, since it comes from a highly reliable source and reports confirmed information. Conversely, data with an 'E5' or 'F6' status would represent the other extreme: the source is unreliable (E) or its reliability cannot be judged (F), and it is either highly improbable that the information provided is accurate (5), or the accuracy cannot be judged (6).

These kinds of qualifications can provide important background information when reporting on the results on the different dimensions constituting a context or stakeholder analysis.

From trustworthy sources to reliable estimates

Once trustworthy sources have been identified, the analyst can start with working towards eliciting reliable estimates. The following two techniques are useful to do so.

Triangulation

A key principle of empirical work – whether it is done by an investigative journalist, a judge, a scientist or a humanitarian worker – consists of the independent verification of facts. Facts that have been confirmed by more than one trustworthy

Table 4.1 Source reliability rating matrix

Source rating scale	*Source trustworthiness rating*	*History of trustworthiness*	*Source authenticity*	*Source objectivity*	*Source access to information*	*Source not vulnerable to manipulation*	*Meets number of criteria*
A	Reliable	Yes	Yes	Yes	Yes	Yes	All 5
B	Usually reliable	Yes	Yes or No	Yes or No	Yes or No	Yes or No	History, plus 3
C	Fairly reliable	Yes	Yes or No	Yes or No	Yes or No	Yes or No	History, plus 2
D	Not usually reliable	Yes or No	Yes or No	Yes or No	Yes or No	Yes or No	2 of 5
E	Unreliable	No	Yes or No	Yes or No	Yes or No	Yes or No	n.a.
F	Cannot be judged	No basis for evaluating the reliability of the source					

Source: based on work of the Canadian Intelligence as described by Hibbs-Pherson and Pherson (2013: 100).

Notes

To be rated ‘A’, the source authenticity must be verifiable, and the source’s history of reliability must be verifiable post-factum or by independent verifiable means. To be rated ‘A’ or ‘B’, the source’s reporting must always be reliable with no significant errors. To be rated ‘C’, the majority of the source’s reporting must be accurate and actionable. To be rated ‘D’, a minority of the source’s reporting must still be accurate and actionable.

Table 4.2 Data accuracy rating matrix

Data rating scale	*Information accuracy rating*	*Independent verifiable means*	*Source subject competency*	*Logical*	*Practical and plausible*	*Consistent*	*Meets number of criteria*
1	Confirmed	Yes	Yes	Yes	Yes	Yes	All 5
2	Probably true	No	Yes	Yes	Yes	Yes	4
3	Possibly true	No	Yes or No	Yes or No	Yes or No	Yes or No	3
4	Doubtfully true	No	Yes or No	Yes or No	Yes or No	Yes or No	2
5	Improbable	No	Yes or No	Yes or No	Yes or No	Yes or No	0 or 1
6	Cannot be judged	No basis for evaluating the validity of the information					

Source: based on work of the Canadian Intelligence as described by Hibbs-Pherson and Pherson (2013: 100).

source are more credible and reliable than facts based on only one source. Establishing the reliability of an estimate by cross-checking it through different sources is called triangulation. Triangulation is even stronger if there is also variation in the type of sources. For example, evidence for the severity of a food crisis would be very strong if it is independently confirmed, e.g. by (a) doctors reporting systematic underfeeding of a large proportion of their patients, (b) a household survey including assessments of caloric intake, conducted by a medical anthropologist working in the field, (c) reports in the local newspaper.

Triangulation attempts usually face two major challenges. First, it will not always be possible to verify the reliability of an estimate. In this case, the analyst needs to decide whether or not the estimate should be included. This may be warranted if the source has a high level of trustworthiness, and there are no alternative estimates available for other indicators covering the sub-dimension. Second, different sources may yield different – and sometimes even inconsistent or contradictory – estimates. In this case, the analyst should first try to resolve the inconsistencies by investigating what might have caused them. For example, the estimate of source A may refer to a slightly different time period or geographical level of analysis than the estimate of source B. For example, does the information refer to a nation, a region or a sub-group in a city? Similarly, estimates based on statistical evidence may be based on different algorithms, or on different definitions of the analytical categories. For example, who classifies as 'unemployed' in official statistics can vary across countries and through time. If this step does not resolve the inconsistency, and the sources are all considered to be equally trustworthy, the analyst may decide to use both of them, e.g. by using their average, and letting the highest and the lowest estimate define the upper and the lower boundary of the confidence interval (see below).

Confidence intervals

Confidence intervals provide interval estimates by defining an upper and a lower boundary for an estimate. The interval spans the range of values that is likely to contain the unknown 'true' value. Interval estimates differ from point estimates, which consist of a single specific value. The size of the confidence interval indicates the reliability of the estimate: the smaller the range, the more reliable the estimate is believed to be. Working with confidence intervals is particularly important in situations where estimates are likely to be inaccurate. This is likely to be the case for most humanitarian settings. We will introduce a method of how to use confidence intervals in measuring and visualization in Chapter 5.

Conclusion

The quality of any context analysis stands and falls with the quality of the evidence on which it is based. This chapter described the most fundamental steps that have to be taken to ensure a decent quality of collected evidence and the estimates that follow from it. Collecting valid and reliable data on emergency settings of course is not an easy endeavor: there are usually never enough time, staff and other resources available for inquiry and fact checking. However, as this chapter has tried to make clear, this does not mean that some minimum standards of data collection and management have to be violated.

The steps outlined in this chapter force the analyst to be explicit not only about whether or not the evidence meets these minimal criteria, but also – in case the analyst decides to proceed – to provide good reasons why such a violation is considered acceptable, and which precautions are taken to prevent mis- or over-interpretation of the evidence. Finally, it is important to realize that – even in a situation of trustworthy sources and good data quality – the numerical estimates may create the illusion of a precision of measurement that does not correspond with the subjective nature of most judgments that have to be made when conducting a context analysis of emergency settings.

5 Conducting a Comprehensive Context Analysis (CCA)

Andrej Zwitter

This chapter demonstrates the application of the security studies-based approach to context analysis. This approach results in what we call a *Comprehensive Context Analysis* (CCA). It gives a picture of the level of security of the various context dimensions at a given time in a decomposed way ready for interpretation. This chapter presents visualization and interpretation strategies for the CCA. Visualization is an important part of context analysis for several reasons. First, good visual aids allow reducing large amounts of data, thereby facilitating interpretation. Second, figures and graphs can have much value added for effective communication, because they structure the available information. Third, though any data-reduction step always bears the risk of oversimplification, graphical representations force the analyst to explicate the assumptions underlying the analysis, thereby facilitating a critical dialogue with the audience. For that purpose we have developed a set of tools for visualization and analysis. The chapter first presents the main tool, the *radar graphs*. The second section expands on how to interpret radar charts.

Measurement of context dimensions

The measurement of threats to contexts is complex. When looking at the basis of the assessment of the CCA, the individual context dimensions (see the following chapters), one realizes quickly that not every dimension of security relevant to the individual can easily be measured with quantitative data. This is due to the fact that some of the threats, which a human being could be exposed to, are mainly of a perceptive nature; other threats to the individual are simply not assessable in a quantitative manner; and again for other threats we simply lack any data whatsoever. Due to the nature of humanitarian action, the necessary data is not always available in the assessed case and the analyst has to rely on proxy indicators, by which we mean an indirect measure based on a theoretical link. Most importantly, the assessment of the security level of the context strongly depends on the indicators the analyst based his research on and on the evaluation of the indicators fulfilling one of the descriptors. Nevertheless, the attempt should be made as precisely and transparently as possible.

Scaling of context dimensions

The scaling presented here makes use of a table (see Table 5.1) that contains a descriptor for every security level in each context. The descriptors combine indicators for threats and trends. The kind of ranking is absolute as there are no competing variables to check. Most importantly, the assessment of the security level of the context strongly depends on the indicators the analyst based his research on and the evaluation of the indicators fulfilling one of the descriptors. It can therefore be somewhat prone to biases. Pre-established indicator lists (see Chapters 6–10) and scaling based on voting in teams can help to compensate for these biases.

This approach to scaling relies much on the evaluation of qualitative and quantitative data. Indicator lists can serve as an aid when making the first steps towards a CCA. However, their use requires a constant and critical assessment of the appropriateness of the indicators used, as well as an open mind to other potentially more important points of attention. For example, be aware that indicators do not always mean statistical data. One can also think of qualitative indicators of a specific context that are just as valuable or even more valuable than quantitative indicators. A potential disadvantage of quantitative data is that the level of aggregation can be quite high (i.e. it refers to large groups of people), so that it might become difficult to identify small, vulnerable groups or particular patterns. Hence, be open to both qualitative and quantitative indicators. Also be aware that indicators are rarely perfectly 'objective' measurements of a situation and can also be related to perceptions or subject to certain biases.

Table 5.1 Contextual security levels and descriptors

	Level	*Descriptor*
Security	A	Ideal state of security as defined; or state close to ideal with no significant indication of threats to a security context.
Latent threat	B1	Average or just below average state of security; some vague indications of threats, which in total are no threat to an aspect of security; no indicators for negative trends.
	B2	Below average state of security; some indications of threats, which in total are no threat to an aspect of security in the long run; accompanied by indicators showing negative trends.
Manifest threat	C1	Single indicators or a combination of indicators, which in total threaten to cause damage to a population in the long run.
	C2	Single indicators or a combination of indicators, which in total threaten to cause immediate damage to a population.
Acute threat	D1	Single indicators or a combination of indicators, which already cause systemic damage to a population.
	D2	Single indicators or a combination of indicators, which already cause widespread damage to a population or result in a high number of victims.

The next step is to translate quantitative data as well as qualitative data into a grid that allows an assessment of a comprehensive situation. In order to do that, it is useful to rely on a table of security levels that provides descriptors for each level in each context. Such descriptors need to relate to the indicators used and the results they have produced. One can differentiate between indicators that show that a certain context is already under pressure (death counts, infection rates, levels of malnutrition, etc.) and indicators that relate more to processes that show a certain trend (demographic changes, increase in hate crimes). These indicators are captured in the descriptors in Table 5.1. The first three levels of security (A, B1, B2) relate to the definitions of theoretically ideal security states and averages. These descriptors require the analyst to make an evaluation of the threat indicators and their impact on the context. Furthermore, they ask the analyst to look for trend indicators specifically, in order not to overlook small facts with potentially big explanatory value. In general, A–B2 are structured in the following way:

- relation to the definition of an ideal condition of context;
- threat indicators;
- trend indicators.

The last four levels (C1, C2, D1, D2) focus more on the clear presence of threat indicators and their estimated effect over time or in terms of impact. The difference between levels C and D is that level C concerns threat indicators that show either immediate or long-term consequences and level D concerns indicators that show a count of context related victims (deaths). Level D1 refers to systemic damage – systemic means that victims are not coincidental but the result of a systemic weakness that when it remains unaddressed would result in widespread damage (D2). It thus follows such a structure:

- threat or victims indicators;
- quality of damage: time, structural embeddedness, spread.

The sub-chapters on the contexts provide examples for high and low levels of security and give some more detailed guidance on the specific aggregation and scaling.

Where exact indicators cannot be found for a case study a less precise indicator or a proxy-indicator can be chosen. A proxy-indicator is an indirect measure based on a theoretical link.

Using confidence intervals

More often than not it will be difficult to pinpoint the exact threat level of a context dimension. As already explained in the previous chapter and illustrated by Table 4.1, this can have several reasons. It can be the result of a low validity of the data due to the fact that the analyst had to use proxy-indicators. Uncertainty about the exact level can also be the result of a low reliability of the data (e.g. coming from an unreliable source). Therefore, we suggest making use of a confidence interval. This interval indicates how confident one can be about the specific assessment of a context dimension. For example, having much confidence in the validity and reliability of data in the health context, because the data on numbers of infections (high validity) was gathered by the World Health Organization and was corroborated by field staff (high reliability), one would choose a very narrow interval (e.g. B1–B2, Table 5.2) or even no interval at all (e.g. B1–B1 or C2–C2, Table 5.2). Having little confidence in the reliability of sources, because field staff has reported that the number of infections seems to be higher than indicated by the data obtained from the WHO, would lead to broadening the interval (e.g. B1–C1 or C2–D2, Table 5.2). If both the source and the data are not very reliable then one would have to opt for an even broader confidence interval (e.g. B1–C2, Table 5.2). As a general rule one can say that the bigger the confidence interval, the less one can rely on the context dimension to be a potential asset in the project design. Table 5.2 should be read together with Table 5.1.

When visualizing the context, applying the confidence interval to the visualization results in a banner (see Figure 5.1: the confidence interval is the dark grey space). This banner indicates in what state the analyst judges the context dimension to be and how much one can build on that context dimension as a resource for the project design. In Figure 5.1 one can see that the analyst's confidence in the assessment of the economic dimension is very high (0) whereas confidence in the assessment of the political dimension is on the lower end (2).

Visualization and interpretation of the Comprehensive Context Analysis[1]

In order to assess the security state of the different context dimensions (food, health, environment etc.), they are visualized in a so-called *radar graph*. Radar graphs have the advantage that they allow (1) to illustrate the interconnectedness between the different contexts, (2) to simultaneously present the situation before and after the crisis and (3) to indicate the confidence intervals related to the

Table 5.2 Confidence intervals

Confidence	*Data reliability*	*Source reliability*	*Interval steps*
High 1	High	High	0 (e.g. B1–B1)
High 2	High	Medium	0–1 (e.g. B1–B1 or B1–B2)
High 3	Medium	High	0–1 (e.g. B1–B1 or B1–B2)
Medium 1	Medium	Medium	1 (e.g. B1–B2)
Medium 2	High	Low	1–2 (e.g. B1–B2 or B1–C1)
Medium 3	Low	High	1–2 (e.g. B1–B2 or B1–C1)
Low 1	Low	Medium	2 (e.g. B1–C1)
Low 2	Medium	Low	2 (e.g. B1–C1)
Low 3	Low	Low	3 (e.g. B1–C2)

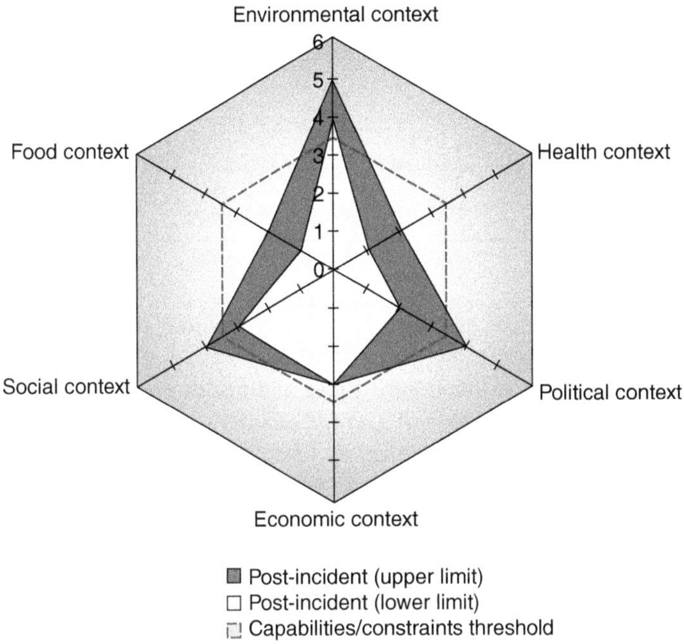

Figure 5.1 Post-emergency mapping of the six context dimensions.

estimate – all in one graph. Since a radar graph can be seen as the unique standardized fingerprint of a crisis, it also facilitates comparison across different crises.

In order to compile such a radar graph for the CCA, the estimates of the higher and lower end of the confidence interval for each context dimension need to be mapped into values between 0 and 6 on the scale of security levels shown in Table 5.3 (also see Table 5.1).

The outcome for the higher end and lower end confidence interval (e.g. environmental security B2–B1=4–5) for each dimension is entered into a spreadsheet (Table 5.4), which is further on visualized in the radar graph (see Figure 5.1).

Table 5.3 Scale of security levels

Threat level	*Value*
A (security)	6
B1 (latent threat)	5
B2 (latent threat)	4
C1 (manifest threat)	3
C2 (manifest threat)	2
D1 (acute threat)	1
D2 (acute threat)	0

Table 5.4 Comprehensive context security levels after a hypothetical disaster event

Context dimension	*Post-incident (lower limit)*	*Post-incident (upper limit)*
Environment	4	5
Health	1	2
Political	2	4
Economic	3	3
Soc.-Cultural	3	4
Food	1	2

Figure 5.1 would be the result of such a CCA mapping after the event of a food crisis. It shows which contexts are under pressure (social, political and economic dimensions) or even in extreme peril (food and health dimensions) and which aspects are in the acceptable zone of average security (environmental).

This method of visualization allows grasping basic patterns of the six contexts in one graph. Note that neither the order in which the radar graph presents the different context dimensions, nor the lines between them, have any specific analytical value. In the example, the environmental dimension (B2–B1) is relatively stable, whereas the health and food dimensions (both C2–D1) are very low. This pattern suggests the following interpretation. First, we could conclude that this might predominantly be a food crisis. Second, since also the health context ranks low, both food and health problems seem to be interrelated. At least, this would deserve some further investigation.

Interpretation strategies for the Comprehensive Context Analysis

The previous step provided a quick descriptive overview into the major problems in the target population. It is necessary for subsequent analytical steps, in which at least the following questions have to be answered:

- What is the current security level in the different context dimensions?
- What are the major problems and constraints in the different context dimensions?

The most straightforward tool we discuss here is the *pre-post-incident analysis*.

Pre-post-incident analysis: working with baseline data

A humanitarian crisis is the product of vulnerability and hazard (Twigg 2004, see Chapter 2). It is, however, difficult to distinguish between the impact of the hazard and the impact of the vulnerability to the hazard on the state of the overall context when only analyzing the situation after the disaster. In order to be better able to identify the vulnerable context dimensions, i.e. those most affected by a hazard, comparing the states before and after the hazard is a useful approach. The purpose of a pre-post-incident analysis is to compare the situation before and after the disaster (as is also proposed in the MIRA method) and helps identifying specifically vulnerable dimensions. This is especially useful in crises with clear-cut disaster events, but can also be done for more enduring crisis events (such as prolonged conflict) by assessing the context dimensions for varying time points and then identify trends in terms of an improving or decreasing crisis situation.

A context analysis always gains in analytical power if the radar chart incorporates information on the situation before the emergency. The reason is that this allows assessing the relative impact of the disaster in terms of a deterioration or improvement of the status quo *ante*.

Figure 5.2 and Table 5.5 now include the baseline dataset (with confidence intervals) and illustrate the CCA mapping of the hypothetical scenario before and after the hazard occurred. In Figure 5.2 the outer banner represents the situation before the food crisis. This we take as baseline data. The inner banner represents the security states in the multiple dimensions after the crisis.

A comparison of the pre- and post-emergency values reveals that the population faced security threats already before the emergency (in the domains of social and economic security). Even though the emergency had its strongest effects on the domains of health and food security, much attention should be paid to social and economic security. A further conclusion from the comparison is that due to the fact that the disaster had a sizable impact also on the political stability, this dimension might have been vulnerable already before the hazard struck, albeit the indicators might not have been so clearly visible to the observer. For reconstruction and disaster risk reduction efforts it would be interesting to investigate whether it was the food crisis, which caused such a drastic decline in the domain of health security (deterioration from B2–B1 to D1–C2), and, if so, how. For example, there may be structural reasons, like the enduring social and political exclusion of the affected group, which might lead to an unequal access to food and medicines. Social inequality may also be the reason why the emergency has not resulted in a decline of security in the social-cultural dimension (which was already relatively low to begin with). Finally, it could be investigated to what degree the decline of health contributed to the decrease of

Table 5.5 Scaling of pre- and post-incident scaling

Context dimension	*Pre-incident (lower limit)*	*Pre-incident (upper limit)*	*Post-incident (lower limit)*	*Post-incident (upper limit)*
Environment	5	6	4	5
Health	4	5	1	2
Political	4	5	2	4
Economic	3	4	3	3
Soc.-Cultural	3	4	3	4
Food	4	5	1	2

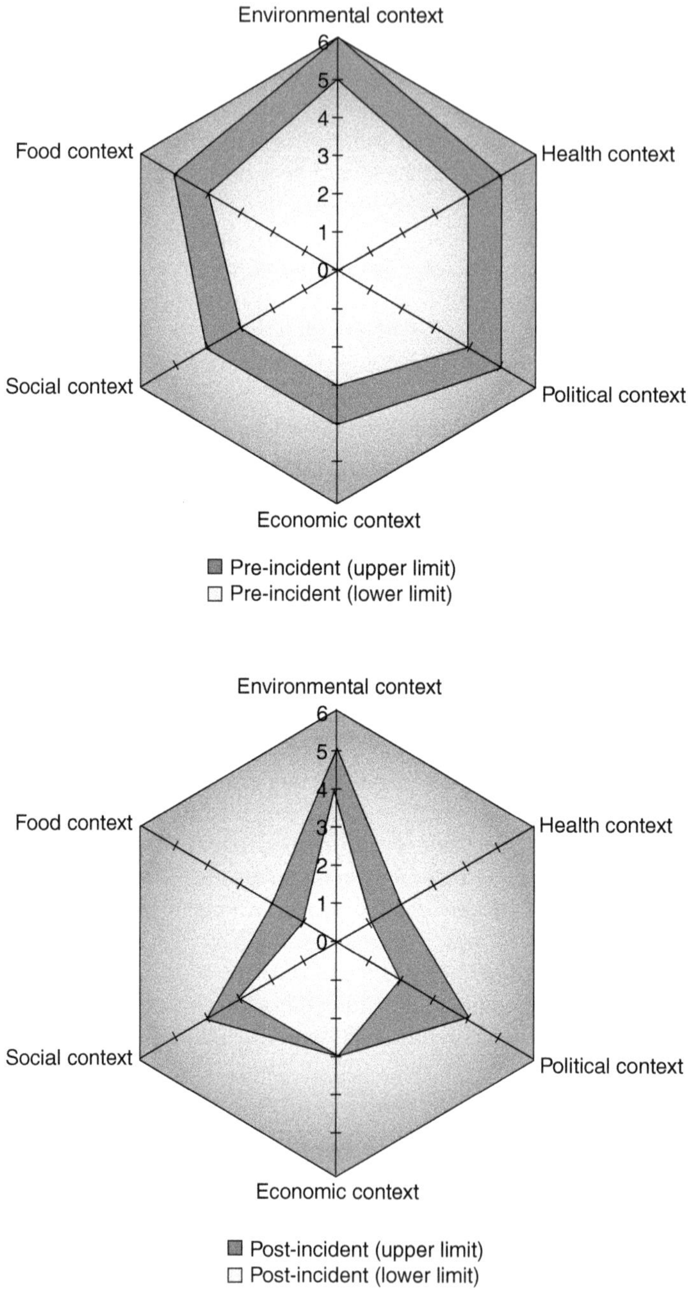

Figure 5.2 CCA mapping before and after a hypothetical emergency.

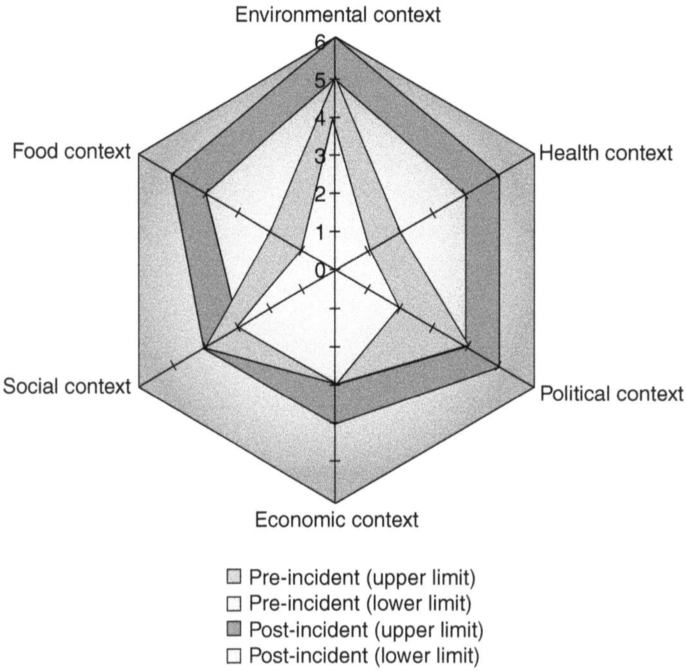

Figure 5.2 Continued

economic security – ill people have a lower productive capacity. In sum, the radar graph shows potential patterns and developments, which generate new questions that require further investigation.

Conclusion

This exploration does not reflect hard numbers. As always the numbers merely represent the aggregate analytic effort of the analyst in condensed form. The Comprehensive Context Analysis (CCA) provides an opportunity to visualize potentials and constraints or established relationships between the contexts before unknown or overlooked. Of course also actors, who are often also identified in a context analysis, influence how the different context dimensions look. This information becomes even more important in Chapter 12 where we will explicate how certain actors can become stakeholders once certain intervention options have been formulated. We will also present further tools for visualizing and analyzing the stakeholder field around particular intervention issues. Hence this first overview conducted through the CCA does not yet lead to a full understanding of the crisis and why relationships exist between various context dimensions. But it gives very good insights into the basic features, such as the capabilities and constraints.

The following Chapters 6–10 will delve deeper into the six context dimensions the H-AID framework uses to conduct the CCA. Each of the chapters will briefly explore the disciplinary and theoretical background of the respective context dimension, suggest common indicators and introduce ways to aggregate and scale data (see Table 5.3).

Note

1 This book features CCA software tools as a web application and an Excel file at http://ecfar.org/CCAnalysis.html. These software tools were developed by Andrej Zwitter.

6 The political context

Christopher K. Lamont

Both for students and practitioners of humanitarian assistance a firm grasp of the political dimension of a context in which a humanitarian emergency occurs, constitutes a crucial component of context analysis. Indeed, Mike O'Neill, Director of Save the Children, noted that every aid worker should not only be aware of his or her particular mission or organization, but also how this mission 'is likely to be perceived by stakeholders in the area of operation' (Mertus 2009: 169–70). While this chapter does not focus on stakeholder analysis as such (which is discussed in Chapter 12), it implicitly highlights the importance of the political dimension for better understanding domestic and international stakeholder perceptions with regards to operational security. Only a firm grasp of the political dimension will inform the design of appropriate, safe and effective humanitarian and disaster response programs. This understanding is important because these perceptions can fuel local responses to disaster relief that may render assistance ineffective or at worst threaten the security of humanitarians in the field. For example, as the Director of Operations for the International Committee of the Red Cross, Pierre Krahenbuhl, notes, the incorporation of humanitarian action into governmental counterinsurgency strategies, as observed in Afghanistan, can have devastating effects upon humanitarians in the field: 'When humanitarian action becomes part of strategies aimed at defeating an enemy, the risks for aid agencies in the field grow exponentially' (Krahenbuhl 2011).

In order to provide the reader with a practical guide to political context analysis, this chapter will first introduce the subject before moving on to providing a brief overview of key definitions. Then operationalization and indicators will be discussed. This will be followed by a discussion of aggregation and weighting. Examples will be provided that will elucidate lessons imparted from this chapter.

Prior to a discussion of history and definitions it is important to point out two inherent risks to the student and practitioner of humanitarian action when assessing the political context of a particular complex emergency. These risks have often resulted in either non-actionable reporting or even, more perniciously, reports that fail to provide decision-makers with an accurate and timely snapshot of a political environment in which an organization seeks to operate. The first of these risks is a failure to explicitly set out what constitutes political context so as to avoid delimiting our discussion of politics to include a broad range of related issue areas such as the social or the economic dimensions of a context. The second is the risk of seeing political context as static and unchanging. Political contexts are always fluid, and therefore require constant revaluation and monitoring of indicators.

History

Traditionally, political context analysis has been the realm of state institutions: foreign ministries, intelligence agencies, defense ministries; however, given the growing number of non-governmental actors engaged in humanitarian relief there has been a concomitant decentralization of political context analysis. This is not surprising, given different actors operate within different risk environments and have distinct vulnerabilities. The forthcoming sections of this chapter should be read with both of the above in mind.

The political context in which interventions take place requires those engaged in humanitarian action to understand the political dynamics in the territories in which an intervention is planned. To be sure, establishing the political context of an intervention can be perceived as a daunting task for the humanitarian analyst as complex emergencies often strike regions or territories suffering from breakdowns in governance or outright civil conflict. An example of this would be the December 2004 tsunami, which struck the conflict-affected province of Aceh in Indonesia. The humanitarian response to this disaster ultimately brought once warring parties into a peace process that resulted in the 2005 Helsinki Memorandum of Understanding. In such cases, often those who exercise authority over a given territory are not traditional state actors, but rather militias, tribal authorities, semi-autonomous local government entities or warlords. Furthermore, public institutions are often not able to provide access or security in a disaster affected region or the state may itself attempt to impede humanitarian delivery (Zwitter and Lamont 2014).

Conversely, humanitarian emergencies also strike states where public institutions are not under conflict related stress, such as Hurricane Katrina in the United States or the Fukushima disaster in Japan. Indeed, as Heintze and Zwitter succinctly observed: ‘however advanced a country’s social, economic and legal system may be, disasters are always a challenge’ (2011: 140). Indeed, in such settings political context indicators may attest to a high degree of political stability, yet at the same time the state may find itself unable to provide relief or security in the hours, days or weeks following a humanitarian emergency. Thus, it is important to distinguish political context analysis from organizational security analysis. While political context often informs operational security, there is not necessarily a correlation.

Thus, given the fact that humanitarian emergencies often entail at least a short-term breakdown in the ability of the affected state to provide relief or security, such as in the context of the United States in the aftermath of Hurricane Katrina, in contexts where there are pre-existing conflicts, or humanitarian emergencies generated by conflict itself, an understanding of political context, political security and political risk is an essential antecedent to intervention.

Political context assessment is often associated with a wide range of more long-term conflict, post-conflict or post-disaster programs that include state-building, rule of law or justice sector programs (OHCHR 2006), and security sector trainings. Many of these programs are often presented as capacity-building (see for example UNDP Afghanistan Country Office 2008). Yet, as seen in Aceh, the interconnectedness between natural disaster and political conflict can witness a blurring of the lines between natural and man-made disaster response. This case requires an acute understanding of the political context, an intractable decades-old secessionist conflict, in the midst of a humanitarian emergency triggered by a natural disaster, the December 2004 tsunami. This interconnectedness provided a window of opportunity to both respond to the natural disaster and engage hitherto belligerent parties in a dialogue that eventually brought about the 2005 Helsinki Memorandum Agreement.

Definitions

For the purposes of clarity, *political context* will be defined as an overarching assessment of public authorities and actors, which exercise legal and/or extra-legal power over a given territory, either through the use of public institutions or through the use of force. Meanwhile, *political security* will be defined in terms of the presence, or absence, of public institutions that exercise legitimate authority over a given territory. *Political risk*, on the other hand, refers simply to the likelihood of political change altering the outcome of a given intervention.

Operationalization and indicators

In the area of political context there are a number of indicators and indices that attempt to provide researchers and practitioners with a snapshot of state political instability. One caveat is necessary to mention here. Often these indices report data on a state-by-state basis, and quite frequently there can be significant variations within states in relation to the indicators included within these indices.

The Fund for Peace's Failed States Index includes six political and military indicators: state legitimacy, public services, human rights and rule of law, security apparatus, factionalized elites and external intervention (Fund for Peace

Table 6.1 Key definitions

Political context	Overarching capacity of public authorities and actors, which exercise legal and/or extra-legal power and policy-making processes within a given territory, either through the use of public institutions or through the use of force.
Political security	The presence, or absence, of public institutions that exercise legitimate authority over a given territory.
Political risk	The likelihood of political change altering the outcome of a given intervention.

http://ffp.statesindex.org/indicators accessed March 16, 2014). The Economist Intelligence Unit also has its own index of political instability, the Political Instability Index (EIU 2014a). The Political Instability Index ranks states on a scale of 0 to 10, with 0 representing no insecurity and 10 representing maximum insecurity. Like the Failed States Index, the Political Instability Index draws from a broad range of political and economic indicators. Based on research conducted by the Political Instability Task Force (PITF) based at George Mason University, this index looks (besides economic indicators as proxies for state performance) at the underlying political vulnerabilities (EIU 2014b) shown in Table 6.2.

In a more simplified manner, Mertus (2009: 171) identifies a number of factors that should be included within a political risk assessment. While Mertus' list is in some respects overly broad in the sense that natural disasters and economic factors are also included, eight factors are of relevance to our discussion within this chapter: history of communal violence and state-sanctioned violence, political elites, external players, existence and functioning of legal institutions in line with the rule of law, protection of human rights, military developments and influence in decision-making, infrastructure, and prevalence of crime. These factors can be used to group indicators that are presented in this selection of potential indicators (Table 6.3).

When thinking about political context, it is helpful to group indicators into core elements. Table 6.4 sets out these core groups and provides suggestions as to methods and techniques that can be deployed by the student or practitioner.

Aggregation and weighing

Indicators provide a means to monitor a given political context, but in order to provide concise guidance to decision-makers these indicators must be scaled. In relation to political security, the scaling as provided in Table 5.1 (Chapter 5) can be used. It allows the analyst to describe varying states of political security

Table 6.2 Political vulnerabilities in the Political Instability Index of the Economist Intelligence Unit

1	Inequality
2	State history
3	Corruption
4	Ethnic fragmentation
5	Trust in institutions
6	Status of minorities
7	History of political instability
8	Proclivity to labor unrest
9	Level of social provision
10	A country's neighborhood
11	Regime type
12	Regime type and factionalism

Table 6.3 List of potential indicators for measuring the political dimension

Level of confidence in key state institutions as expressed in opinion polls
Citizen's perception of the state and state institutions as measured through focus group-based scenario drawing
Assessment of the state and state institutions through contents analysis of key media and professional journals
Perceived level of corruption (opinion polls)
Data on number of cases of corruption uncovered and classification according to the type of corruption
Number of electoral processes, and the level of voter participation
Number of political parties with members in parliament
Number of major daily newspapers and radio and TV stations independent of government
Representation of minorities or disadvantaged groups in legislatures, the judiciary and the executive arms of the government
Legal provisions in respect of human rights specifically to safeguard minorities or disadvantaged groups
Degree of ratification of international human rights treaties, covenants and conventions
Modification of the constitution and the law in conformity with the ratified conventions
Revising administrative procedures (e.g. Ombudsman)
Administration of justice indicators (e.g. court cases heard, trials, sentences, prisons, type of crimes, arrests and detentions, number of cases brought against state institutions)
Attracting public support through civil society, the media, schools, etc.
Number and type of cases submitted to the Ombudsman institution (if in place)
Political refugees and IDPs
Press freedom index
Human rights and freedoms index

Table 6.4 Core groups and suggested methods

Institutions	
Perception of legitimacy of state institutions	Open source information
Corruption	Open source information; existing rankings (e.g. Transparency International)
Political participation	
Political exclusion and inclusion; presence of significant excluded groups (racial, ethnic, religious, political)	Open source information (banned political parties, exclusionary laws, discriminatory practices)

over time. For example, Japan's Fukushima prefecture prior to the triple catastrophe of earthquake, tsunami and nuclear disaster would have rated A for optimal state of security; however, in the aftermath of the March 11, 2011 events, the same geographic region would have received a C2, denoting a significant breakdown in the ability of the state to exercise authority in the worse affected communities and the risks posed by the evolving nuclear disaster at the Fukushima nuclear plant.

The weighing of indicators and the scaling within the political dimension requires a look at multiple elements, as the Haiti earthquake illustrates. Haiti's governing institutions have been characterized by despotism and chronic brutality nurtured in the context of a complex geopolitical environment in the Western Hemisphere as would be suggested by indicators such as regime type, corruption or political violence. The result was a failure to foster democratic institutions (Khouri-Padova 2004, Muggah 2009). Like many developing countries struggling with poverty and instability, Haiti's high unemployment, almost non-existent institutions and crippling infrastructure all contributed to a deteriorating security environment and by extension to a very problematic political environment. However, it also is important to point out that Haiti lacks the ethnic identity-based divisions that fuel instability elsewhere (Collier 2009). These last points illustrate that the political dimension cannot be viewed separately but must be always combined with insights of the other context dimensions.

When the January 12, 2010 earthquake struck Haiti, it occurred in a country that was already experiencing acute political instability. With a magnitude of 7.3 on the Richter scale, and an epicenter in close proximity to the capital Port-au-Prince, the earthquake devastated the political heart of Haiti. As such, the calamity of natural disaster was coupled with a breakdown in governing institutions in an already weak state.

The capacity to respond to the earthquake was severely limited as the Haitian government lost 33 percent of employees and the UN lost 102 staff members (IASC 2012a: 51). The Post Disaster Needs Assessment (PDNA) conducted by the World Bank and other donors evaluated the damages and losses at US$8 billion or 120 percent of GDP (World Bank 2012: 6). As such, in addition to suffering a major humanitarian catastrophe, Haiti – already a weak state – lost its ability to effectively govern its own territory. Taking as a reference point of scaling Table 5.1 (Chapter 5), one can see that Haiti was dealing with a combination of indicators, which caused systemic damage to the population. This leads us to conclude that at least the level of 'Acute Threat D1' was reached. When looking whether the indicators also show widespread damage to the population as a result of political instability related violence (rather than of the direct health, food, social and economic consequences) it becomes clear that the political dimension has not totally eroded to cause widespread loss of life such as for example encountered in a genocide. Therefore, given the above, even in the absence of ethnic exclusion or armed insurgency, Haiti would score in the *Acute Threat* range of *D1* in the aftermath of the 2010 earthquake due to the loss of governing institutions in the context of an already weak state.

This rating would have been easily attainable through background research that made use of initial media reports emerging from Haiti on the scale of the disaster and the extent to which it impacted Haiti's capital Port-au-Prince. From existing reports that pre-date the earthquake, it would be apparent that Haiti lacked functioning governing institutions. Therefore, already, we can start from an assumption that the earthquake, which devastated Haiti's capital, would have decimated governing institutions.[1] The relief operation that was launched in the aftermath of the earthquake took this into account in the sense that much aid was delivered with the assistance of the US military, which provided a secure environment in which humanitarian aid deliveries could take place. In the absence of strong local interlocutors, external actors largely dictated the immediate response to the emergency and thus, in a political security environment such as Haiti, one can expect to meet with low levels of political resistance.

In conclusion, this chapter and its examples are instructive of the importance of taking into account political security considerations in humanitarian planning. The scaling and indicators provided constitute a template that will allow for humanitarian practitioners to more effectively communicate political developments on the ground to humanitarian decision-makers.

Note

1 '25% of civil servants in Port au Prince died; 60% of government and administrative buildings, 80% of schools in Port-au-Prince and 60% of schools in the South and West Departments were destroyed or damaged' (DEC 2014).

7 The economic context

Fleur S. Mulder and Bartjan J.W. Pennink

Introduction

In this chapter we will focus on how to capture the economic situation in a specific region prior, during or after a humanitarian crisis, if one is not able to do very extensive research and a quick overview of the economic situation is needed. We will do this by describing a set of dimensions related to economic development on a national, regional and local level. This overview will then be part of the overall Comprehensive Context Analysis that can be created after reading and applying all context chapters of this book.

The above-mentioned levels do not reflect a normative view on what the defining level of analysis should be, or which level is most important, as elaborated previously in this book. It simply reflects the potential levels of analysis for H-AID. In one aspect, however, we think that there is a difference in importance. In the field of humanitarian crises we think that, especially in terms of short-term recuperation, the local level is the most important. The local level is a combination of the group level and the individual level. By focusing on the local level we think the most relevant information can be gathered to cover the domain of the socio-economic world in times of humanitarian crises. However, most of these dimensions can also be assessed on the regional or national level, as we will show later in this chapter.

Our locally focused list can serve as a basis for governments and aid organizations to identify where and how local, regional and national economies could be stimulated. For example, governments can set up subsidiary programs and aid organizations can initiate livelihoods projects. All this can help in supporting, restoring and developing economies after crisis has hit.

For each level we develop a limited list of economic dimensions and indicators. This list will differ from general lists of dimensions for economic development, because often general lists only show the state of the economy at a national level. We will start with the local level and then continue with the other two levels. With regard to the indicators, low(er) scores on the proposed indicators imply conditions that are relatively undesirable: less business, more unemployment, more inflation, less infrastructure, less roads, more electricity cuts, etc.

History and definitions

In this section, we will present local dimensions of the economic context, followed by an additional set of regional and national dimensions. We will discuss why these are important dimensions of economic development and growth and, when relevant, what their potential relationship to other dimensions is. This theoretical part will function as a basis for the next section, in which we will address how to measure each dimension.

Local dimensions of the economic context

Below we present a set of local dimensions of the economic context. These dimensions can also often be analyzed at the regional and national level.

Poverty

In 1995, at the World Summit on Social Development in Copenhagen, scientists defined extreme poverty as 'a condition characterized by severe deprivation of basic human needs, including food, safe drinking water, sanitation facilities, health, shelter, education and information' (Alina *et al.* 2010: 209). In 2004, the European Council presented a relative definition of poverty:

> [b]ecause of their poverty they may experience multiple disadvantages through unemployment, low income, poor housing, inadequate health care and barriers to lifelong learning, culture, sport and recreation. They are often excluded and marginalized from participating in activities (economic, social and cultural) that are the norm for other people and their access to fundamental rights may be restricted.
>
> (Alina *et al.* 2010: 209)

Poverty can be analyzed by focusing on wage and income, and poverty standards. Regarding the first (wage and income), economic theory holds that 'at the aggregate level the growth of real wages is determined by labour productivity growth' (Sharpe *et al.* 2008: 1). If wages rise in line with labor productivity they are both sustainable and create a stimulus for further economic growth through individuals' purchasing power increase (Cristescu *et al.* 2014). Factors that may explain a weak connection between the growth of wage and productivity might be price and wage rigidities, labor adjustment costs, employment protection, restrictions on entry to the labor market and other market regulations (Klein 2012). Labor adjustment costs are costs that are caused directly by adjustments in, for example, size and division of the workforce within a company. As Cahuc and Zylberberg (2001) point out, the size of such costs is often considerable, which is why they play an important part in decisions to hire or fire staff. Employment protection is a set of mandatory restrictions governing the firing of employees. Its purpose is the increase of the volume and stability of employment (Cahuc and Zylberberg 2001).

Regarding poverty standards, Alina *et al.* (2010) state that income or consumption expenditures are the only variable representing the welfare level of an individual. A large study conducted by Meyer and Sullivan (2012), however, presents strong evidence that a well-constructed consumption-based poverty measure would be preferable to income-based measures of poverty for determining the most disadvantaged. Still,

> while a consumption-based measure of poverty may be used to set overall standards for program eligibility, individual consumption data are not suitable for determining eligibility for antipoverty programs. Given that at least some components of income, such as formal earnings and transfer income, are easier to collect and validate, income will typically be more appropriate for determining program eligibility for individuals or families.
>
> (Meyer and Sullivan 2012: 111)

Accessibility of basic goods

The process of economic development is related to the economy's ability to satisfy basic physical contributors to the quality of life (Knowles 1997). Welfare and productivity are related, 'as improvements in basic needs fulfilment enhance people's strength, agility and stamina, which can stimulate productivity' (Perlo-Freeman and Webber 2009: 965). Higher levels of productivity in turn generate a greater pool of resources from which investments in basic needs fulfillment could potentially be drawn, making the relationship between accessibility of basic goods and economic development a circular one. However, the direction of causality seems to run more strongly from basic needs fulfillment to economic growth than the other way round, with reverse causality only clearly apparent in the case of nutrition (Perlo-Freeman and Webber 2009).

Employment

Employment growth is a key indicator of labor market performance (Clayton *et al.* 2013). High levels of employment are a sign of high productivity, high income (relative to a situation with little to no employment), high purchasing power (which further stimulates the economy), reduction of poverty, lower crime levels and thus of an overall blooming economy and society (Inekwe 2013).

As will be explained later in the paragraph about small businesses and entrepreneurship, in times of recession and crisis people who lose their jobs may be pushed into self-employment. When an economy in recession shows growing numbers of entrepreneurship, this can mean that employment rates are actually lower than they seem, due to 'hidden unemployment' within this group of recently self-employed. Hidden unemployment means that people might stay employed but for lower wages than original and/or needed to survive. In the Western context we evaluate this as not efficient and, therefore, negative. In other contexts, however, this evaluation can be different: instead of doing what has to be done with as few employees as possible in order to make labor as cheap as possible, it can also be a consideration to divide one job over several people so that more people gain an income out of it. The income for all will thus be lower, but more people have at least something.

Unemployment cannot be studied independently from wages (Zhu *et al.* 2009). A large number of empirical studies have shown that there is a negative relationship between the (regional) real wage and the unemployment rate, usually called the wage curve (Blanchflower and Oswald 1994).

Transportation infrastructure

Transport infrastructure can function as a means to bring about territorial cohesion, reduce economic disparities and promote economic development (Crescenzi and Rodríguez-Pose 2012). Furthermore, transport infrastructure improvement affects firms' demand for labor and therefore also affects regional (un)employment levels. It is important here to distinguish between the effects on voluntary and involuntary unemployment. Indirect welfare effects of infrastructure improvements may only arise through changes in involuntary unemployment, whereas these effects through changes in voluntary unemployment are absent (Zhu *et al.* 2009).

The transport sector is related to all other economic activities (Usón *et al.* 2011). A high quality infrastructure shortens distances between production and consumption centers, responds to the mobility demands of citizens and ensures good accessibility within a territory. 'The close correlation between the economic development of a region or country (measured in terms of GDP) and transport (measured in travelers or tons per kilometer) is, therefore, not surprising' (Usón *et al.* 2011: 1917).

A minimum threshold for transport infrastructure is necessary. This level does often not yet exist in developing countries. Once such a threshold has been achieved, however, there seems to be little evidence that infrastructure endowment and new infrastructure investment can become a catalyst for sustainable economic growth (Crescenzi and Rodríguez-Pose 2012).

Small businesses and entrepreneurship

For several decades, it has been thought that small businesses are the foundation of job growth. This thinking is backed up by data from the Business Employment Dynamics (BED) program at the Bureau of Labor Statistics (BLS). These data show that firms with fewer than 500 employees – the criteria often used for defining small firms – account for about two-thirds of net jobs created (Clayton *et al.* 2013).

The problem with targeting young, small businesses as the focus of job creation is that the outcomes of new businesses are diverse. To focus on those businesses that are truly job creators, economists and policy-makers are now talking about 'high-growth firms'[1] (Clayton *et al.* 2013). However, high-growth firms form but a fraction of the total of (small) businesses, whereas the remaining set of small businesses form the vast majority of businesses. It is this majority that is the basis for not only a functioning economy, but also for a functioning society. Small businesses help provide a society's primary needs through all kinds of goods and services. Thus, the professionalism within the small businesses segment in any economy says a lot about the quality of life within that economy's society.

Entrepreneurship is related to unemployment in two ways. As unemployment increases, individuals may choose to become self-employed instead of remaining unemployed. As a result, self-employment (entrepreneurship) will increase as individuals are 'pushed' into entrepreneurship. However, the relationship between unemployment and entrepreneurship may also be present in the other direction; as unemployment increases, the overall market may go into decline. This effect reduces the demand for goods and services, resulting in business failures and fewer entrepreneurs. Thus individuals are 'pulled' out of entrepreneurship when unemployment increases (Gohmann and Fernandez 2013). From a policy perspective, public policies that reduce the costs of entrepreneurship and ease the transition from unemployment into proprietorship can potentially have positive short- and long-run impacts on unemployment (Gohmann and Fernandez 2013).

Currency

Analyzing the currency of a country or region is important for humanitarian organizations, because by entering a local situation and trying to find out what has to be done it makes sense to pay attention to the way people do business with each other. First of all, one needs to know whether there is any form of business life at all. Second, one should deduct whether any existing business is constituted of an exchange of goods (trading) or whether the economy is money based. Third, the value of eventual money in relation to the exchange and/or trading of goods should be determined.

MANNER AND EASE OF TRADING

The functioning level of an economy is closely related to the degree of monetization: one should first conclude whether an economy is money based or trade based. When it is money based, both the degree of monetization within an economy and the liquidity of an eventual stock market are of importance in researching the economy's viability (Levine 1997, Tennant and Abdulkadri 2010).

INFLATION PERCENTAGE

The value and stability of the currency used within an economy are of great importance. The inflation percentage greatly influences both. The relationship between inflation and economic growth has been subject of discussion among economists. Some argue that there is a negative relationship between inflation and economic growth and claim that high inflation rates lead to a decrease in output, which subsequently leads to low economic growth in the long run (Odhiambo 2013). Others argue that it is the variability of the inflation rate, and not the level of inflation, that negatively affects growth. Then there are those who argue that there is a threshold below which a moderate increase in inflation promotes growth, but above which further increases in the inflation rate actually retard economic growth. This threshold, however, differs from country to country and over time, and is thus hard to establish. Odhiambo (2013) concludes that there is a bidirectional causality between inflation and economic growth, meaning that inflation and economic growth directly influence and even cause each other.

Energy efficiency

Liimatainen and Pöllänen (2013: 150) defined energy efficiency as the ratio between total haulage and energy consumption, indicated as tonne-kilometers per kilowatt-hours (t km/kW h). Although energy use and emissions increase when economic development takes place, so does an economy's energy efficiency (Liimatainen and Pöllänen 2013). The Environmental Kuznets Curve (EKC) refers to the idea that policy-makers within any given economy can only then actively embed sustainable development into their plans when there is a situation in which the economy's average income reaches a certain point and the majority of the country's inhabitants no longer lives in poverty (Shafik 1994). In other words, if the first priority is mere survival, one cannot afford to invest in their surroundings until their personal situation is of a certain quality level. In developing economies, pollution and consequential health concerns initially increase as industrialization rises, due to a lack of both financial and practical means to prevent these negative side effects (Tierney 2009). Therefore, sustainable development is a phenomenon typically seen in more advanced economies.

Innovation capacity

Within business life, innovation capacity is determined by, among other things, savings mobilization and risk diversification (Tennant and Abdulkadri 2010).[2] An increased effectiveness in savings mobilization will release more funds for investment and thus increase the possibilities for innovation and economic growth. Regarding risk diversification, a uniform distribution of loans across all sectors is assumed to be the highest level of portfolio diversification. A low level of sectoral diversification would, in turn, negatively impact on the ability of financial institutions to mobilize further savings (Tennant and Abdulkadri 2010).

Additional economic context dimensions on the regional and national level

The above-mentioned dimensions can be measured at the local, regional and national level, but in times of humanitarian crises we deem it most relevant to analyze these dimensions on the local level. Below we present a set of additional economic context dimensions at the regional and national level, which will be of more importance to assess once the immediate crisis phase has passed.

Large businesses (regional level)

It is generally believed that especially large businesses and multinationals can create spillovers, indirectly stimulating economic growth (Che and Wang 2013). The channels through which multinationals can affect economic growth include 'productivity gains, human capital turnover from foreign companies to domestic firms, the diffusion of technology, and the creation of a competitive environment to "push" domestic firms to work harder' (Che and Wang 2013: 1). However, the ability of a country to take advantage of multinationals' spillovers appears to be dependent on local conditions.

The effects of multinationals on economic growth are stronger in countries with relatively well-developed financial markets (Alfaro *et al.* 2004). The presence of multinationals has both a significant and a positive effect on economic growth only when these economies have a good property rights protection, which relates this dimension to financial legislation (Che and Wang 2013). The potential positive influence of large businesses and multinationals is important for all economies. However, for developing or recovering economies in particular, it is important to increase their attractiveness to large businesses by creating an environment that is beneficial for their economic health, such as a low taxes economy.

Foreign Direct Investment (national level)

Foreign Direct Investment inflow to agriculture in developing or recovering economies is very desirable because employment of unskilled and skilled labor and national welfare unequivocally improve by FDI in agriculture (Inekwe 2013, Chaudhuri and Banerjee 2010).

This dimension links to the dimension of financial institutions as will be described below (Qayyum *et al.* 2014). It is in the presence of sound economic institutions that foreign inflows with clear development or recuperation agendas can be utilized efficiently and contribute positively, while a system with weak institutions may result in moral hazard[3] and rent-seeking[4] problems that will reduce the productivity of capital. Still, in general, foreign aid accelerates economic activity and generates tax revenues that enable governments to focus on enforcing rules of law by fighting corruption effectively (Qayyum *et al.* 2014).

Financial institutions (regional and national level)

The existing economic literature clearly suggests that financial systems perform five basic functions which aid in the creation of economic growth (Tennant and Abdulkadri 2010). As summarized by Levine (1997), these functions comprise:

(1) savings mobilization; (2) risk diversification; (3) efficient allocation of resources through acquisition of information about investment opportunities; (4) exertion of corporate control after projects have been financed; (5) facilitation of exchange in the economy. There is a growing consensus that financial sector development can lead to economic growth by means of financial sector deepening and sophistication through liberalization (McKinnon 1973 and Shaw 1973 in Tennant and Abdulkadri 2010). These functions can thus not be effectively performed in isolation of each other.

Within any financial system, three types of financial institutions can be distinguished that play a role in the execution of the above functions. First, there are depositary institutions, such as banks, credit unions, mortgage and loan companies and trust companies. Second, there are contractual institutions, like insurance companies and pension funds. Last, there are investment institutions, such as investment banks, brokerage firms and underwriters.

Until recently, little attention had been paid to a comparison of the relative effects of different types of financial institutions on economic growth. Recent research by Tennant and Abdulkadri (2010) suggests that models which isolate the impact of any one type of financial institution on economic growth do not yield good results, while estimations that implicitly account for the contribution of all types of financial institutions generally perform better.

Financial legislation and enforcement (national level)

A direct and positive correlation exists between an economy's quality of governance and the height of per capita growth rates (Qayyum *et al.* 2014). Good governance enables a country to achieve its development goals and become prosperous by establishing a conducive environment for high and sustainable economic growth (North 1990, 1992). Good governance includes the establishment of impartial, predictable and consistently enforced rules.

One goal of financial legislation is consumer protection. Another is corporate control. Financial institutions can perform this corporate control function by monitoring managers of firms who have been allocated credit or investment funds. As connected party loans and/or connected party financial investments increase, the ability of financial institutions to exert corporate control decreases (Tennant and Abdulkadri 2010).

Education and literacy (national level)

Better educated people are more flexible to adapt to technological innovations and more likely to invent; typically, they are more productive, earn higher wages, are more geographically mobile and seem to cluster in high-growth industries (Strauss and Thomas 1998 in Perlo-Freeman and Webber 2009). However, both within and across countries, inequalities in educational opportunities exist. Variations in the rates of return to education across countries are due to three factors: quality of education, externalities of education and inequalities in education (Perlo-Freeman and Webber 2009). The quality of education matters and is largely determined by externalities like the amount of money a government is willing or able to spend on education and educational material (textbooks, computers, teachers etc.). Inequalities within developing countries create asymmetries in the ability to pay to attend fee-paying schools. Even when schooling is free in the poorest countries, child labor is often pervasive and can be a necessity for many families, constraining children's educational attainment (Amin *et al.* 2006). A lack of availability of tertiary education – identified as having the greatest economic impact on output (Knowles 1997) – and also of secondary and sometimes primary education may be the reason why economic growth is stunted in some less developed countries (Perlo-Freeman and Webber 2009).

National budget and resources (national level)

The national budget and available resources are also of importance for a country's economy. By assessing the budget deficit and the resource flow one can get more insight in this national dimension of the economic context.

BUDGET DEFICIT

One factor that can severely constrain both the ability of developing and recovering countries to meet the basic needs of their populations and their economic growth is unsustainable foreign debt burdens (Perlo-Freeman and Webber 2009), caused for example through the reception of humanitarian aid. Many developing countries built up such foreign debts during the 1970s and 1980s. High (foreign) debt is widely seen by NGOs, governments, international organizations and academics as causing widespread human suffering and economic damage (Perlo-Freeman and Webber 2009). IMF-imposed programs to enable countries to maintain repayments have often involved the removal of food subsidies and the introduction of user fees for health and education services, creating a direct link with basic needs fulfillment. Conversely, there is evidence that global efforts to write off or reduce the debt of the poorest countries under the Heavily Indebted Poor Countries (HIPC) initiative have led to genuine welfare gains. Greenhill and Blackmore (2002), for example, show that debt relief has led to large increases in health and education spending in Africa (Perlo-Freeman and Webber 2009).

RESOURCE FLOWS

Resource allocation can have a positive effect on economic growth, as it is argued that when credit to private sector production increases as a proportion of total loans, resources are being allocated to the projects most likely to be productive (Tennant and Abdulkadri 2010). Credit allocated to public sector entities, the financial sector and consumption is less productive than credit allocated to other sectors of the economy, such as agriculture, mining and manufacturing, construction and land development, transportation, storage and communication, electricity, gas and water distribution, tourism and entertainment, and professional and other services sectors (Tennant and Abdulkadri 2010).

Operationalization and indicators

In this section, we will elaborate upon ways to measure the dimensions described above. We first discuss the indicators of the local dimensions that are considered core to an analysis of the economic context prior, during or after a humanitarian crisis. Table 7.1 summarizes the dimensions and the associated indicators. As said, these dimensions can also be assessed on the regional and national level, so we have incorporated information about these levels as well. The indicators will all be accompanied by recommended methods of measurement that, if necessary, are specified per level.

In cases in which it is not possible to find 'hard' data upon which an indicator can be rated, we suggest finding stakeholders or experts who are familiar with issues regarding that particular local indicator. One can ask them to provide their professional opinion and give a score for each of the indicators. To partly overcome the subjectivity of their perceptions, always more than one stakeholder should be questioned, with an ideal number of four to eight stakeholders.

A simple, yet pragmatic and unbiased method of measuring the stakeholders' opinions is to ask each of them to give an impression of the state of a particular indicator and to mark the idealness of the situation on a continuum of 10 centimeters with both ends labeled as the extremes from low to high. Afterwards, the researcher can determine the score by measuring the distance from the ends of the continuum; in the preferable case of having more than one stakeholder, the mean can be derived from all measurements.

Aggregation and weighing

Once one has collected data regarding the above-mentioned dimensions and indicators, one needs to establish whether the situation is worrisome or not. The higher the score on one of the indicators, the more economic activity or development there will be. In other words: in situations in which a large-scale disaster took place, we expect the scores for most indicators to be low. In line with the other chapters we will also need a second translation from all scores on all indicators to one score as a whole on one dimension: the economic context. This score could range from hardly any economic activity to a lot of economic activities.

As economic situations and needs vary between locations, it is impossible to define universal thresholds for the above-mentioned indicators. As we explained previously, the opinions of the local stakeholders will be decisive in the scaling and valuation process. Since researchers using this quick scan will be obtaining much of their data (especially their local data, which is most the important data) from verbal sources using a scaling system from 1 to 10, we recommend to use the same scaling system throughout the quick scan model: 1 being the ultimate

Table 7.1 Core dimensions of the economic context

Local economic context dimensions	*Indicators*	*Data sources*
Poverty	Income Consumption	Local and regional: statistical data provided by official institutions; stakeholders. National: statistical data provided by official institutions.
Accessibility of basic goods	Clean drinking water Food Sanitary facilities Gasoline Health care	All levels: statistical data provided by official institutions; stakeholders. See also Chapter 9 on the health context.
Employment	Employment rates	Local and regional: statistical data provided by official institutions; stakeholders. National: statistical data provided by official institutions.
	Hidden unemployment	All levels: stakeholders.
	Wages	Local and regional: statistical data provided by official institutions; stakeholders. National: statistical data provided by official institutions.
Transportation infrastructure	Number of (damaged) kilometers of motorways, railways Number of airports Number of travelers per kilometer on a yearly basis Tons of goods transported per kilometer on a yearly basis	Local and regional: statistical data provided by official institutions; stakeholders. National: statistical data provided by official institutions.
Small businesses	Number of small businesses Professionalism of small businesses Number of high growth businesses Measure of hidden unemployment among entrepreneurs	Local and regional: statistical data provided by official institutions; stakeholders. National: statistical data provided by official institutions. All levels: stakeholders.

Currency	The degree of monetization	Local and regional: statistical data provided by official institutions; stakeholders.
	Inflation percentage	National: statistical data provided by official institutions.
	Overall currency stability	All levels: statistical data provided by official institutions.
		All levels: statistical data provided by official institutions; stakeholders.
Innovation capacity	Savings mobilization[1]	All levels: statistical data provided by official institutions; stakeholders.
	Risk diversification[2]	
Energy efficiency	Measure of pollution	Local and regional: statistical data provided by official institutions; stakeholders.
	Measure of infrastructure (see results indicator 5)	National: statistical data provided by official institutions.
	Energy input to GDP	National: statistical data provided by official institutions.

Notes

1 This can be measured for each type of financial institution by the total deposits of the institution, divided by the total assets less the loans of the institution, i.e. SMOB = Deposits/(Total assets – Loans). This ratio measures the effectiveness of financial institutions in using the resources at their disposal to attract savings (Tennant and Abdulkadri 2010).

2 For each type of financial institution this is calculated by first finding the percentage of total loans allocated by sector. The SD of the percentage of total loans allocated to each sector is then used to measure the spread for each institution from the state of uniform distribution. If all loans are concentrated in one sector the maximum SD is derived; however, if loans are allocated equally to each sector, then no spread around the mean exists, and the minimum SD possible equals zero (Tennant and Abdulkadri 2010).

Table 7.2 Additional dimensions of the economic context

Additional economic context dimensions	*Indicators*	*Data sources*
Large businesses	Number of large businesses and multinationals Professionalism of business life Number of large high-growth businesses	Regional: statistical data provided by official institutions; stakeholders. National: statistical data provided by official institutions. Regional and national: stakeholders.
Foreign Direct Investment	Amount of money coming from FDI FDI income as a percentage of national income Amount of FDI per sector	National: statistical data provided by official institutions.
Financial institutions	Number and professionalism of depositary institutions Number and professionalism of contractual institutions Number and professionalism of investment institutions Ability of the financial system to perform its five basic functions (Levine 1997) 1 Savings mobilization 2 Risk diversification 3 Efficient allocation of resources 4 Exertion of corporate control 5 Facilitation of exchange	Regional and national: statistical data provided by official institutions; stakeholders. Regional and national: stakeholders.
Financial legislation and its enforcement	Implementation of law and order Level of corruption Quality of bureaucracy	National: statistical data provided by official institutions; stakeholders. See Chapter 6 on the political context.
Education and literacy	Quality of education and educational material General education level (primary, secondary or tertiary) Literacy rates Women's education	National: statistical data provided by official institutions; stakeholders. National: statistical data provided by official institutions.
National budget and resources	Foreign debt to GDP Resource allocation	National: statistical data provided by official institutions. National: statistical data provided by official institutions; stakeholders.

low point, 10 being the best possible situation and 5 representing a situation that is just enough. In this scaling system, every value below 5 would deserve special attention and could be labeled as insufficient.

In humanitarian work it is desirable to not only have the scores of each of the indicators after the crisis or disaster but also the scores on the indicators before the disaster or crisis. This gives the opportunity to give a better interpretation of the seriousness of the situation related to the situation before the disaster. If the data is not available these can be replaced by asking some key stakeholders to construct the scores based on perceptions for the local and even perhaps the regional level. For the national level data can often be found in archives and databases.

Aggregating local level indicators

This level incorporates the following eight dimensions:

1 *Small businesses and entrepreneurship.* A low number of small businesses, a low score of professionalism in small businesses, a low number of small high-growth businesses and a high score of hidden unemployment among entrepreneurs will lead to a low score on this indicator.
2 *Employment rates.* Low employment rates, a high score of hidden unemployment and a low average wage will lead to a low score on this indicator.
3 *State of transportation infrastructure.* A low number of kilometers of (accessible) motorways and railways, a low number of airports, a low number of travelers and a low number of transported goods will lead to a low score on this indicator.
4 *The presence or absence of poverty.* A low average income and high levels of consumption in comparison to income will lead to a low score.
5 *Currency/basis of trade.* A low degree of monetization, a high inflation percentage and a low measure of overall currency stability will lead to a low score on this indicator.
6 *Energy efficiency.* A high score of pollution, a low quality of infrastructure and a negative correlation between energy input and GDP will lead to a low score on this indicator.
7 *Innovation capacity.* A low score of savings mobilization and a low score of risk diversification will lead to a low score on this indicator.
8 *Accessibility of basic goods.* Poor accessibility of clean drinking water, poor accessibility of basic foods, poor sanitary facilities, poor accessibility of gasoline and poor quality and accessibility of health care will lead to a low score on this indicator.

Aggregating regional economic context indicators

This level incorporates the following nine dimensions, of which the indicators of the following dimensions have already been discussed in the previous paragraph: (1) small businesses and entrepreneurship; (2) employment rates; (3) the presence or absence of poverty; (4) the state of transportation infrastructure; (5) currency/basis of trade; (6) energy efficiency; (7) innovation capacity.

To these are added indicators for (8) *large businesses*: a low number of large businesses and multinationals, a low score of professional life and a low number of large high-growth businesses will lead to a low score. (9) *Number, professionalism and stability of financial institutions*: a low number of depositary institutions, a low score of professionalism of depositary institutions, a low number of contractual institutions, a low score of professionalism of contractual institutions, a low number of investment institutions, a low score of professionalism of investment institutions and a limited ability of the financial system to perform its five basic functions will lead to a low score.

Aggregating national economic context indicators

This level incorporates the following 13 dimensions, of which the following have been discussed above: (1) small businesses and entrepreneurship; (2) large businesses; (3) employment rates; (4) the presence or absence of poverty; (5) the state of transportation infrastructure; (6) currency/basis of trade; (7) number, professionalism and stability of financial institutions; (8) energy efficiency; (9) innovation capacity.

To these are added indicators for (10) *Financial legislation and its enforcement*: a low implementation of law and order, a high level of corruption and a low quality of bureaucracy will lead to a low score. (11) *Educational level and literacy rates*: a low quality of education (or educational material), a low general level of education, low literacy rates and a low score of women's education will lead to a low score. (12) *National budget and resource*: a positive and high correlation between foreign debt and GDP and inefficient or lacking resource allocation will lead to a low score. (13) *FDI*: a low absolute amount of money coming from FDI, a low amount of FDI as a percentage of national income and a low amount of FDI per sector will lead to a low score.

Overall aggregation

To construct a one-dimensional state of the economic situation we have to bring back the scores on the above set of indicators to one overall score. However, it might be really difficult to obtain hard objective data at the local level, so it might be good idea to use confidence intervals (see also Chapter 4). With the help of confidence intervals one can present the most negative estimation and the least negative estimation of the economic situation, based on the impressions of key stakeholders. With this it will be possible to make an overall score with an upper and lower bound value.

The first step is however to aggregate the scores on the set of indicators. For this purpose we choose for different weights of the three levels. For humanitarian crisis and disaster management the local level is the most important level for economic context analysis, as argued previously. Hence, we suggest to construct for each level a mean of the scores (i.e. a mean for the local level, a mean for the regional level and a mean for the national level). We suggest that within each level all the indicators will have the same weight. The second step is to weigh the local level three times and the other two levels (regional and national) one time for the overall score. We assume here that all the information has been translated into quantitative data; if exact information is not available estimations have to be made.

For example, if for the local situation we measure a score of 2, meaning that there is not much economic activity, and for the regional level a score of 4 and at the national level a score of 5, then the final overall score will be derived as follows: $0.6 \times$ local score $+ 0.2 \times$ regional score $+ 0.2 \times$ national score $=$ final score for the economic context ($0.6 \times 2 + 0.2 \times 4 + 0.2 \times 5 = \ldots$). As one can see, the local situation has a dominant influence on the final score.

Notes

1 High-growth firms are a very small subset of all firms. However, they contribute substantially to job creation. Definitions of high-growth firms vary, but Clayton *et al.* (2013: 2) state that the period for defining high-growth firms should be long enough such that short-run transitory changes in employment are not falsely measured as high growth.

2 Savings are of great importance in the light of economic development. In times of need, such as shortly after any type of crisis, the mobilization of savings enables parties to smooth spending, investment and consumption.

3 A so-called 'moral hazard' means that a situation is facilitated in which a party may develop a tendency to take risks because the costs (be they financial or other in nature) that could result will not be felt by themselves.

4 Rent-seeking is spend wealth in order to increase one's share of existing wealth without creating new wealth.

8 The social-cultural context

Cecile W.J. de Milliano and Barbara Boudewijnse

Introduction

> Disasters are interesting because they tell us something about human behavior. The disaster calls upon communities to respond, and exploring these community responses provides information about how a community functions, treats outsiders, and rises to new challenges. Community responses also inform us about culture, the habitual ways of doing things, and the meaning ascribed to these patterns of action, about what is 'normal' and 'abnormal' behavior in a given community, and of how this can be stretched to its outer limits by the circumstances.
>
> (Ride and Bretherton 2011: 7)

Through thoughts, actions, feelings, decisions and interactions, humans give shape to their everyday social life because 'human beings are the architects of their own social reality' (Newman 2004: xxii). Their social and cultural background influence the way they shape this reality. Moreover, humans have psychological and social needs (e.g. sense of belonging, affection, confirmation, status). The social-cultural context allows one to realize these needs. Yet, if humans are deprived of fulfilling these needs, they suffer. For example, loneliness (deprivation of social relations) has been found to be a stronger predictor of premature death than smoking (Holt-Lunstad *et al.* 2010). In crisis situations, people not only have social-cultural needs, they are also often brought together with others who they are normally not in contact with. In the latter case this for example includes local communities, national and international aid organizations, armies, government officials, media. People's social-cultural background influences how they deal with crisis situations. Unfortunately, the importance of understanding this dimension is often neglected.

In other words, an important element of a context analysis is assessing to what degree the means to fulfill basic socio-cultural needs are present in a certain context. This chapter focuses on several concepts to get an understanding of the social-cultural context of daily life. In addition, it offers concepts and tools to identify social-cultural patterns of a society. The chapter starts with an introduction to five common concepts in the field of social behavioral sciences. These key concepts cover a part of the socio-cultural context and include: *social identity, social structure, social control, stratification* and *social change*. The second section discusses how to apply these concepts as part of a context analysis of (complex) emergencies. We have formulated a set of key questions per key concept as summarized in Table 8.1, below. The third section explores how to distinguish between situations of 'high' and 'low' social security.

Definitions: five basic concepts

When coming to grips with the socio-cultural context of 'daily life', the first question that arises is 'of whose daily life?'. In this chapter the focus is on the daily life of a society which refers to 'a population living in the same geographic area that share a culture and common identity and whose members are subject to the same political authority' (Newman 2004: 19). Exploring 'daily life' of societies (i.e. those dealing with crises) through socio-cultural glasses means exploring their social reality. Social reality, also referred to as culture, is what a group of humans experience as 'the common way of life'. Sociologists and anthropologists believe that members of societies see certain facets of daily life in similar ways because of culture which provides them with a common bond and shared meaning (Bauman 1991, Howard 1989, Cohen and Kennedy 2000, Steuer 2003, Lekha Sriram *et al.* 2009, Eriksen *et al.* 2010, Newman 2004: 115). Culture is a result of the complex pattern of living that humans develop and pass on from generation to generation (Adler and Adler 2007: 433). Language, values, beliefs, rules, behaviors and artifacts characterize the human experience of their social reality (Newman 2004: 587). Culture influences how members of societies acquire a (collective) sense of self, how values, norms and ideals are instilled, how reality and truths are constructed and how social order is created and maintained. Therefore, insight into a society's culture is particularly important during times of crises.

The following sections will explain five concepts which enable getting insight in the social-cultural patterns of a society. In each section, the concepts will be applied to the context of the earthquake in Haiti. The *Goudougoudou* – as the Haitians call the earthquake of January 2010 – killed around 230,000 people and disrupted the lives of 1.3 million (Bellegarde-Smith 2011). Its epicenter was close to the most densely populated areas of Haiti (including the capital Port-au-Prince). The case study allows understanding the social-cultural context in daily reality.

Social identity

The first concept, *social identity*, contains essential aspects of a society, which consists of humans' sense of self, ethnicity, race, gender and religion (Newman 2004: 128).[1] Language, traditions, rituals and religions contribute to a social identity. By mastering, for example, a language, performing traditions and rituals, taking roles and responsibility, humans can acquire a sense of self. Social identity is of importance as it can influence the likelihood of communication, cooperation, governance and solidarity. However, normative clashes between groups can also increase in societies which are more ethnically and culturally diverse (Newman 2004: 115). Not mastering or understanding certain powerful characteristics of a group's social identity, such as not speaking the same language, can exacerbate problems in inter-cultural communication and even lead to conflict.

During Cyclone *Nargis* in Myanmar in 2008, for example, language was a key problem for international staff dealing with a region that was rural and had few foreign language speakers. Apart from problems with oral communication, strategic and technical documents, meetings and minutes used for coordinating the international response, were almost all in English and translations were rare. At times this caused frustration and disputes between local and internal emergency response actors (Poussard and Hayter, in: Ride and Bretherton 2011: 158).

Developing a social identity, internalizing 'cultural information' and learning to become a 'functional member' of a society, is a lifelong process (Nolan and Lenski 2006: 361). This is achieved through *socialization*: the process through which one learns how to act according to the rules and expectations of a particular culture (Newman 2004: 129–30). It is a process of learning in which individuals ascertain what the society defines that individuals should and should not do. For example, the roles and responsibilities that men and women take in a society are influenced by gender socialization. Through this process individuals learn to see themselves as a male or female and learn the roles and expectations associated with that sex group (Adler and Adler 2007: 434). Powerful institutional agents of socialization are the family and community structure, formal and informal education, religious institutions and the (mass) media (Newman 2004: 154).

Cultural beliefs have direct effects on the beliefs and behaviors of a society stricken by disaster. For example, to psychologically cope with the recurring flooding in central Java (Pati Regency, Indonesia), the young explained that three cultural values were supportive: *pasrah* – submission to one's fate – *tawakal* – resignation or trust in God – and *nrimo* – acquiescence to one's fate. In times of suffering, the Javanese are expected first to pray, then try to make the situation better, then accept the situation (*nrimo*) and then submit one's fate as given by God/Allah (*pasrah*) (de Milliano 2012: 218). These beliefs are rooted in more general cultural and religious schemas that have been found to be characteristic for Javanese culture, which has its own language, dance, music and mysticism (known as *kebatinan*). For example, most Javanese practice a religion that combines elements of Islam, Hinduism and mysticism (Mulder 1978). According to this schema, a Javanese person should: be striving to maintain harmonious unity (*rukun*); believe that an individual alone is not capable of achieving much; be willing to make sacrifices for the common good; be patient, cooperative and obedient to those with a higher status (Alons 2009, Mulder 1978).

The importance of (not) taking social identity into account can for example be learnt from an interview with a director of a local NGO who responded to the earthquake in Pakistan in 2005:

> So, I think that initial phase brought people together. I think the way the distribution was done, probably tore them apart, yeah. I think the way the distribution was done, we could see the competition for goods which were insufficient to get to everyone ... so in the hysterics of the moment, you probably, you know, grabbed for whatever you could get, but when you came back and you thought and you re-analysed your position and people around you, you know your family and friends and all that, so then you shared once again. And so because the social fabric was so well-knit, I think interventions tried to damage it but couldn't ... So ... er the whole idea ... that notion that when there is a disaster so the best thing for them, for everybody, is to come into camps and get fed and everything, I though, was completely missing the reality of people's lives, yeah?
>
> (Ride and Bretherton 2011: 179)

Social identity in the context of Haiti

Nou fèt pou nou mouri – We are born to die (Haitian saying, Bellegarde-Smith 2011: 265). Haiti has a population of more than nine million people of which almost 60 percent of the population lives in rural areas (Pierre *et al.* 2010). It is a young society, with approximately 50 percent under 20 years of age and about 51 percent of the population being single. In Haitian society, especially in times of stress and difficulties, family ties are of great importance. The family is elastic and extended and usually includes a large network of relatives, neighbors and friends (Pierre *et al.* 2010: 18). Solving personal problems is seen as a family or religious matter. People readily consult elders or religious leaders in their own community (Pierre *et al.* 2010: 25). The concept of the person in Haiti extends beyond Western often individualistic notions of the self and encompasses spiritual dimensions. In the cosmocentric culture of Haiti, the person is seen only as a part of a much larger universe of spirits, ancestors and the natural world, all of which must be in harmony to ensure good health. Haiti is characterized by religious diversity including Roman Catholicism, Vodou (which combines West and Central African traditions and Catholicism) and various Protestant traditions. Each religion cannot be understood without taking the others into account (Pierre *et al.* 2010: 20). Religion plays a crucial role in all spheres of Haitian life, including politics, morals and health.

Social structure

In times of crises, individuals and groups call on their social structures to help them through the difficult times prior, during and after adverse events. Social structure as a concept includes the social institutions, organizations, groups, statuses and roles (Newman 2004: 285–6). Social institutions form a stable set of rules, statuses, groups and organizations that provide a foundation for behavior in major areas of social life (Newman 2004: 28–31). They include for example the institution of education, family, politics, religion, health care or the economy (Adler and Adler 2007: 291). Social organizations are the network of relationships among the members of a society or group (Nolan and Lenski 2006: 361). For the purposes of a context analysis of crisis situations, two dimensions of social structure are particularly useful: formal versus informal social structures, and external versus internal relations as the primary basis of aid.

First, we can distinguish formal and informal social structures. Formal social structures are public or private organizations with a specific division of labor, a hierarchy of authority and a predefined arrangement of statuses and roles. Informal structures refer to small-scale groups. Its members work together not in their official capacities but as persons. Relationships are characterized by face-to-face relations, mutual aid, cooperation and companionship (Sociology Guide 2013). Both formal and informal relationships play a fundamental role in crisis situations, as the case of post-tsunami activities in Aceh (Indonesia) in 2004 demonstrate:

> Village *keuchik* (village heads) or imams and village organizations had a key role in clean-up efforts in many cases. Asrida, volunteer, said there was a sort of 'togetherness' at this time with custom factors influencing togetherness: 'like the keuchik's role, the role of the *ulamas* (religious leaders) to give the spirit back to the people'. (...) So those who didn't use to have any activity, then did some activities with the others (such as religious activities, cooking for women, building houses for men, and young people helping in these activities). Finally we as victims tried to make each other stronger.
>
> (Pelupessy, in: Ride and Bretherton 2011: 31)

Second, we need to distinguish between internal and external mechanisms of aid provision. Internal mechanisms are ongoing and established within the geographical and cultural boundaries of a society (Kieffer 1977). It occurs on three levels (categories) of internal assistance units:

- Primary level (personal), which refers to assistance between people who are well acquainted (family, neighbors and friends). Generally family represents a foundational primary group of society.
- Secondary level (group-community), which refers to assistance in which the recipients may not be known personally by the assisting unit, but whose culture is understood by the assisting unit. This can be assistance between individuals, between groups and individuals and between groups.
- Third level (town-recipient), which refers to assistance, which is less personal but cultural patterns are understood by assisting units.

External mechanisms have their foundations in (and are controlled by) sources external to the recipient community (Kieffer 1977: 3). Two mechanisms are key:

- Those assistance units which are external to the community but internal to the general culture area (referring to the background of the recipient and the assistant unit).
- The foreign assistance unit is characterized by a 'cultural disjunction' between itself and the recipient unit (also referred to as the 'outsider' or 'alien assistance unit'). This category varies from moderate levels (e.g. instanced between ethnicities) to extreme levels (e.g. the assistance and recipient unit have few basic cultural similarities). These levels are determined by the amount of cultural similarities, the length of existence of the unit and finally its functioning prior to a cause or its creation specifically for a circumstance.

Large disaster relief and development programs tend to come from international organizations and from organizations outside the community. Especially in times of severe crises, these external interventions are of importance, since internal mechanisms are often damaged or destroyed. However, by the time external mechanisms reach the disaster sites and are able to intervene, internal interventions are often already actively taking place. Familiarity with and knowledge of the geography, context, culture, language etc. can allow internal assistance units and those familiar to the culture to be very effective. For example, following the *Bam* (Iran) earthquake in December 2003, the local Red Crescent teams pulled 157 people alive from the rubble using far fewer resources than the 34 search-and-rescue teams which flew in from 27 countries and saved 22 lives (IFRC 2004).

Social structure in the context of Haiti

As mentioned previously, the family and its large extended network are of great importance in the Haitian culture. Especially in the urban shantytowns and rural areas, clusters of extended family units share a common courtyard (*lakou*) and form an interdependent community (Pierre *et al.* 2010: 18). Work and child-care are divided among the families sharing the courtyard. Urban families are described as less interdependent and middle class families in urban centers are organized around a model combining Haitian and Anglo-American elements (Pierre *et al.* 2010: 18). The *lakou* is a flexible cultural 'blueprint' underlying crucial forms of coping behavior. It can even be found among internally displaced people living in camps. Nevertheless, due to poverty the *lakou* system has been disrupted, leaving many families without the support and shared parenting afforded by the *lakou* (Pierre *et al.* 2010: 19). If three or more generations are living in the house, age and gender influence power and authority. Generally, the following order is assumed: grandparents, father, mother, the eldest child and so on. In the family, authority is said to be held by the father, who is often absent, thus the mother is the *poto mitan* – the central pillar of the family. In general, mothers have responsibility for the spiritual and emotional life of the family; fathers are responsible for finances. In urban areas, female-headed houses, however, are very common. These women also have to financially provide for their family (Pierre *et al.* 2010: 19).

Social stratification

Interwoven in the fabric of societies is a sense of social ranking and inequality, referred to as stratification. Stratification refers to a process in which groups of people are ranked in a structured system, which perpetuates unequal rewards and life chances in a society. Stratification is linked to *social inequality*, the unequal access to and division of material and non-material goods that are seen as valuable in a society (definition translated from Dutch, Langen 2007: 83).

Stratification is often based on estate and social class systems. *Estate systems* develop when high-status groups own land and have power, which is based on their noble birth (Kerbo 1991 in: Newman 2004: 330) – as is common, for example, in more traditional rural societies in various African countries. *Social class systems* develop when groups of people are ranked based on their income and wealth (Newman 2004: 330). In this system the most common classification is upper, middle and lower (under) class. Additionally, *race*, *ethnicity* and *gender* are important categories, which are used to rank and organize societies. If this results in members of one sex group having more access to wealth, prestige and power than members of the other sex group, this is referred to as gender stratification (Adler and Adler 2007: 432). For example in Burkina Faso, especially among the Mossi and Bisa ethnic groups, the compound functions as the main social unit. It is typically headed by the oldest male in the lineage segment and consists of one or more *households* and *hearth-holds* (Thorsen 2002, INSD 2009). The households include a male head and a number of married and unmarried adult male dependants. The hearth-holds include the wives, daughters-in law, widows and migrants' wives as well as the unmarried children. The compound head possesses considerable symbolic and social power as he performs religious sacrifices to prevent bad things from happening. Moreover, in case of flooding, it is the head of the household who alerts the family, who is the main person in charge and who will be the first to receive information (de Milliano 2012: 208).

Social stratification in the context of Haiti

Haiti is marked by a powerful class hierarchy (Desrosiers and Fleurose 2002), reflected in education, language, economic background and culture. The two official languages are Creole (Kreyòl) and French. However, the latter is only written, spoken and understood by 20 percent of the population, which are usually the elite and middle class urban residents. Valdman argues that: 'the French language has acted primarily as a "social filter" in Haiti, restricting access to spaces of political, economic and social power' (1984 in: Pierre *et al.* 2010: 17). Haiti also has significant social stratification (and discrimination) based on gradations of skin tone, which are part of the legacy of colonization and slavery (Trouillot 1990). This implies that darker skinned people are more likely to be members of lower socio-economic groups and to experience more marginalization. Lighter skinned people are more likely to be members of the elite and of higher socio-economic status (Pierre *et al.* 2010). In addition, there is unequal access to secondary education in Haiti where, although 72 percent of the population has enjoyed primary school education, only 1 percent of the population has a university level education (Pierre *et al.* 2010). Finally, economic background is important as there is a strong divide between the poor majority of the population (95 percent), i.e. the black descendants from the African slaves, and the small minority (5 percent) of the French-speaking elite, consisting of whites and mulattos (descendants from white French fathers and black mothers). Historically, there has been much tension between these two groups. Next to the enormous economic differences between the poor and the elite, color distinctions also play an important role in the perception of social statuses (Pierre *et al.* 2010: 17). People strongly identify with either the black descendants of the slaves or with the mulatto elite. Before the earthquake, Haiti was reinventing itself socially by the integration of social classes (Bellegarde-Smith 2011). However, the earthquake has slowed down the process of allowing all Haitians into the body politic as full citizens and valuable members of society (Bellegarde-Smith 2011: 272).

Social control

Disaster situations can bring both the best and the worst out of people (Ride and Bretherton 2011). Dire need and despair can make people do things they would not normally do, such as stealing or plundering, and also feed aggression and rioting.

Social control is the effort undertaken to help ensure conformity to norms and values (Adler and Adler 2007: 436). The level of social control influences the extent to which people pay attention to each other's behavior, and report on deviant behavior. It is a form of regulating behavior to align with societies' norms and values (Langen 2007: 113). Norms are culturally defined standards or rules of behavior (Newman 2004: 41). They specify what people should do, and what is proper or necessary behavior within particular roles, groups, organizations and institutions.

Sanctions, in the form of reward or punishment, are a means to reach social control and avoid delinquent behavior (Langen 2007: 115). Sanctions can be positive in the form of reward, or negative in the form of punishments. Moreover, there are internal sanctions, which come from within the person (based on values and beliefs such as solidarity and guilt) and external sanctions that are allocated by other members of society. For example, in Myanmar most people relate to each other along family patterns with elders viewed as senior and honored with formal communication. Symbolic acts of dissent, such as not bowing to a military official, can be seen as defiant or threatening. The fear of causing embarrassment or humiliation to another person (*arnade*) strongly influences people's behavior (Poussard and Hayter in: Ride and Bretherton 2011: 144). Another common distinction is between formal and informal sanctions. Social order can be established through legal institutions (formal) or through informal pressure in primary groups. Societies differ with regard to the degree to which both function. But also in emergency settings, the formal legal institutions and their related executive powers (judges, police) may be less operational than usual.

Social control in the context of Haiti

Several factors from Haiti's past, such as the brutal campaigns of the founding leaders, the electoral practice of 'winner takes all' and shifting geopolitical priorities are believed to have consistently prevented the emergence of democratic institutions and have led to overall weak governance in Haiti (Rencoret *et al.* 2010). In addition Haiti's massive unemployment, weak government institutions, lack of public infrastructure and serious environmental degradation are recognized as contributing factors to violence and related security problems throughout the country (Rencoret *et al.* 2010, see also Chapter 6). The security situation in Haiti notably improved in the years preceding the earthquake, however the capacity of the Haitian National Police (HNP) remained fragile and dependent on support from the UN Stabilization Mission in Haiti MINUSTAH and the UN civilian police (UNPOL). Due to the earthquake, like many other branches of government and international actors, the security sector experienced heavy human losses, including 77 dead and 253 severely injured HNP officers (Rencoret *et al.* 2010: 21). In the immediate aftermath of the disaster there were many claims of looting and violence. It is believed, however, that these were largely exaggerated by the international media and did not materialize (OCHA 2010). Nevertheless, HNP and UNPOL representatives expressed concern about a rise in security incidents since the disaster. This included for example an increase in sexual violence (particularly in displaced settlements), social unrest and the recapture of 4,188 inmates that escaped from prisons as a direct or indirect result of the earthquake (Rencoret *et al.* 2010: 21).

Social change

Societies and cultures are far from static and the way in which they alter over time is what is referred to as social change (Adler and Adler 2007: 436). Societies remain stable because enough people define existing conditions as satisfactory and they change because enough people define once-accepted conditions as problems that must be solved (Newman 2004: 527). Social reality is a result of individuals who 'mold society' and society that 'molds the individual'. From one generation to the next, individuals, as members of social groups, acquire their social-cultural way of life. But circumstances change and people have to adapt to these changes. Numerous factors influence and cause social change such as, for example, environmental and population pressure, permanence of settlement, societal complexity, cultural innovation and ideology (Nolan and Lenski 2006: 67, Newman 2004: 508). In addition, because of processes of globalization, people's lives all around the globe, more than ever before, are increasingly interconnected and interdependent. Transnational media, global communication and transportation systems, and massive immigration lead to social change as a result of the worldwide exchange of cultural elements (Newman 2004: 98). In this way, beliefs, technology, customs and other cultural items are spread between groups or societies. When issues arise that concern a large number of people who feel a desire to address these changes, social movements can develop (Newman 2004: 527). For example in various Central American countries, wide community initiated networks around Disaster Risk Management have been developed. These networks became powerful bodies to push national governments for structural change and to strengthen the right to human security and claim their human rights (Morales 2013).

Social change in the context of Haiti

Having faced much historical injustice and the continuous structural violence of global economic policies, many Haitians have learned to maintain hope in the face of severe adversity. Due to the political and economic situation, large numbers of Haitians have migrated to Canada and the United States. Smaller numbers have migrated to other countries, such as France or other Caribbean nations. About 30 percent of Haitian households (about 40 percent in urban areas) have relatives living in high-income countries. The diaspora sends more than US$800 million annually to family and friends in Haiti (Pierre *et al.* 2010: 18). Migration is a major coping strategy, which affects 30 to 40 percent of Haitian households. Economically, it may form a very important source of national income. On the other hand, social structures may be even more disrupted, causing a negative spiral of dependence on migrant money.

Operationalization: key questions

The five concepts presented in this chapter, can be seen as helpful and necessary concepts to understand the social cultural context. Multiple qualitative and quantitative methods exist to explore and understand social-cultural issues and to provide knowledge and information about how societies work and what processes influence societies socially. Examples are field research, surveys, focus groups or experiments (Sarantakos 2005, Lekha Sriram *et al.* 2009, Newman 2004: 82). Although these methods are helpful, it is important to remember that the qualitative aspects of culture are often intangible and thus difficult to measure. They tend to disappear from view when the focus only narrows down to producing *measurable* data or manifest *patterns* (behavior patterns, social 'institutions', etc.). Especially in situations of crises efficiency pressures are high and there is often a need to collect local data as quickly as possible. This, however, holds the danger of producing data that are, so to speak, stripped of their content, context and/or cultural foundations. With this in mind we propose the following 'toolkit' which contains the key concepts and some related questions which are purposefully formulated as open questions. An exploration of these open questions can allow gaining insight in the socio-cultural context. Depending on the available time, the questions can be used in either 'quick scan' or more 'in-depth' methods.

Table 8.1 Key concepts and diagnostic questions describing the socio-cultural context

Concepts	*Questions*
Social identity	With which groups or sub-groups do people identify themselves?
	To what extent can a collective identity be discerned?
	Can different groups with distinctive collective identities be identified?
	What are the cultural expectations (roles and responsibilities) of men and women?
	With what worldview/religions do people identify and to what extent do they play a role?
	With which ethnicities do people identify and to what extent does ethnic identity play a role in inter-ethnic relationships?
	To what extent can either tensions or forms of cooperation between groups with different collective or ethnic identities be found?
	To what extent have norms and values transmitted from one generation to the next?
	How do people share (disaster related) knowledge and expertise?
	What are the expectations in terms of being a 'functional member' of society?
	Who are the role models for a society?
	Do people have access to formal and informal educational systems?
Social structure	What are the key social institutions (organizations, groups, statuses and roles)?
	What are the main networks of (coping) relationships among the members of a society or group?
	Which organizations, groups or networks structure/direct behavior in some major area of social life (education, health, religion, politics etc.)?
	How are communities socially structured?
	How is family life socially structured?
	What are the formal and informal power relationships within and between social groups?

Stratification	What is the distribution of upper, middle and lower class groups (referring to economic position in society based on their income and wealth)? To what extent do status groups own land and have power, which is based on their noble birth? What are the most important material and non-material goods that are seen as valuable in a society? To what extent do the following groups have equal access to material (livelihoods assets) and non-material (i.e. power) goods? • different social classes? • men and women? • different religious groups? • different ethnic or racial groups? • different generations (age-groups)?
Social control	How is social order maintained? What social sanctions (efforts) help to ensure conformity to norms? Positive and/or negative sanctions? Internal and/or external sanctions? What are the formal judicial structures? What are the informal judicial structures? What are the central social norms and values that ought not to be transgressed?
Social change	What is the actual and/or perceived history of the group? What is the impact of technological developments? What is the impact of globalization? What events have had a major impact on collective consciousness? To what extent does migration play an important role? To what extent is tradition seen as important? To what extent is modernization seen as important? What are possible causes of social change? To what extent are the social movements addressing social changes in the society?

Table 8.2 Weighing enabling and disabling categories in the social-cultural context

	Level	*Descriptor*
Optimal enabling environment	A	Ideal state of social security, all five categories score enabling (+5)
Suboptimal enabling environment	B1	Four out of five categories are enabling (+4)
	B2	Three out of five categories are enabling (+3)
Insecurity	C1	Two out of five categories are enabling (+2)
	C2	One out of five categories is enabling (+1)
Disabling social-cultural environment	D1	None of five categories is enabling (0)
	D2	Least ideal state of social security, one or more of five categories are disabling (between –1 and –5)

Aggregation and weighing

Key questions or indicators provide a means to monitor a given social cultural context. By using the information derived from the questions in Table 8.1 one can explore to what extent the social cultural context can have an enabling or disabling impact. The five concepts can either score as being enabling (1 point), disabling (–1) or neutral (0 points). For example, the optimal level of an enabling social-cultural context (which in daily reality is probably an implausible situation) would mean an enabling *social identity*, an enabling *social structure*, enabling levels of *social control*, *stratification* and *social change.* The authors acknowledge this is highly artificial and strongly 'scale' dependent. It must thus be clearly identified if the analysis is taking place on a 'macro' (large scale, national), 'meso' (regional) or 'micro' (local) level. A complicating factor is that often situations are not as clear-cut. It is, therefore, impossible to give a format in which the different categories are weighed in a uniform or absolute procedure. Always, the rating of each of the categories must be related to the scale and be scored in context.

Note

1 The definition of 'identity' is adapted from that used by Newman (2004). 'Ethnicity' is said to refer to a social identity derived from the cultural heritage which is shared by a category of people with common ancestry (Newman 2004: 385). 'Race', on the other hand, refers to a breeding population in which certain traits occur with a frequency that is appreciably different from that of other breeding populations (Nolan and Lenkski 2006: 360). In both political and scientific discourses on 'social identity', however, these conceptual distinctions are blurred. Both concepts are commonly also identified as 'social constructs', as terms used to make social distinctions that have no actual historical or biological basis (see for instance Nagel 1994, Smaje 1997, Gunaratnam 2003, Smedley and Smedley 2012).

9 The health and food context

Rensia R. Bakker

Being healthy is important for each person. Each stage of life asks for specific health needs as well as health services. This includes the protection against infectious diseases and injuries as well as against emerging non-infectious diseases as diabetes, coronary diseases and cancer. Globally estimated for 2008, 60 percent of deaths were attributed to non-infectious diseases and 9 percent due to injuries (due to unintentional accidents as well as to intentional violence). Of the 60 percent who deceased due to non-infectious diseases, only 20 percent occurred in high-income countries (WHO 2011).

Whether one perceives him or herself as being healthy varies per individual and per culture, therefore it is important to define 'health'. For example, in current medical training and literature the definition of 'disease' is mostly based on the biomedical Western perception of health.[1] This can be measured via laboratory tests, for example. Illness is the subjective or behavioral response. Lay people, as well as health care workers, can simultaneously hold both 'traditional' views (based on their culture of origin) and 'biomedical' ones at the same time (Helman 2001).

Food as a means to stay or become healthy is included in our definition of the health context. As providing food is often the core business of humanitarian organizations we introduce this theme more extensively. Furthermore, the food context shows clear links to the other contexts as discussed in the other context chapters.

An example is the link between diarrhea and malnutrition. Water, sanitation[2] and hygiene (WASH) are critical health determinants for survival. In case of inadequate sanitation and/or inadequate water supplies, and inability to maintain good hygiene, the risks are high for outbreaks of diarrheal diseases or health problems caused by vectors associated with solid waste and water, like malaria. Globally, there are 1.7 billion cases of diarrheal disease every year. Diarrheal disease is the second leading cause of death in children under five years old and kills around 760,000 children in this age group each year. Diarrhea is also a leading cause of malnutrition in children under five years old. These children are being trapped in the vicious circle of suffering from malnutrition, which makes them more vulnerable to diarrhea. Each next diarrhea episode thus makes the effect of malnourishment worse (WHO 2013).

History and definitions

In this section we define the central concepts of this chapter.

Health, health security and public health

After the Second World War the entitlement to a healthy life was described in article 25.1 of the Universal Declaration of Human Rights (1948): 'Everyone has the right to a standard of living adequate for the health of himself and of his family, including food, clothing, housing and medical care and necessary social services'.

The most accepted academic definition of health is one of the World Health Organization: 'Health is the state of complete physical, mental and social well-being and not merely the absence of disease or infirmity' (WHO 1948: 2). Critics argued that the WHO definition of health is utopian, inflexible and unrealistic (Saracci 1997). Therefore, in the 1980s, the era of the development of health promotion, the definition was revised into the following:

> The extent to which an individual or group is able to realize aspirations and satisfy needs, and to change or cope with the environment. Health is a resource for everyday life, not the objective of living; it is a positive concept, emphasizing social and personal resources as well as physical capabilities.
>
> (WHO 1984: 20)

The most recent reformulation of the concept of health (security) relates to the protection of health, i.e.

> Health security is at the vital core of human security – and illness, disability, and avoidable death are 'critical pervasive threats' to human security.... Health's instrumental role is collective as well as personal. Good health is a precondition for social stability. Sudden outbreaks of a contagious disease or other health crisis can destabilize an entire society.
>
> (Ogata and Sen 2003: 96)

We note here the need to protect human lives and to provide Public Health services regarding all related needs, which is a much broader definition than the protection against disease alone.

Public Health includes elements of health system organization, economy, political systems, social history and environmental conditions. It can be described as follows:

> Public Health refers to all organized measures (whether public or private) to prevent disease, promote health, and prolong life among the population as a whole. Its activities aim to provide conditions in which people can be healthy and focus on entire populations, not on individual patients or diseases.
>
> (WHO n.d.)

The International Covenant on Economic, Social and Cultural Rights (ICESCR), an international human rights treaty of the United Nations (UN), and the Convention on the Rights of Persons with Disabilities (www.un.org/disabilities/convention/conventionfull.shtml), are legal frameworks that call to protect the health of the inhabitants of a state. Each inhabitant of a nation which ratified the ICESCR should formally be able to rely on its government to provide the basic needs via progressive realization. So, formally, all minorities, internally displaced persons and refugees are secured with the same basic service of health care as the main population, as far as feasible within the financial and organizational limits of the country.

In case of an armed conflict health services are covered via the framework of International Humanitarian Law. In the field of humanitarian action this means that all ill and wounded persons can be taken care of adequately. Medical services, including health workers, medical equipment, hospitals and medical transportation, are all protected under this framework, provided the prescribed requirements are observed, being the principles of Humanitarian Law (Arts 24–29 Convention I 1949, Art. 36 Convention II 1949, Art. 20 Convention IV 1949, Art. 8 Additional Protocol I 1977, Physicians for human rights 2013: 1–4).

Food and nutrition security

Food security was in 2002 defined by the Food and Agriculture Organization (FAO 2002) as follows: 'when all people, at all times, have physical, social and economic access to sufficient, safe and nutritious food that meets their dietary needs and food preferences for an active and healthy life'. The right to food is one of the basic human rights. The right to adequate food is realized when every man, woman and child, alone or in community with others, have the physical and economic access at all times to buy or grow adequate food. Adequate food means enough food in quantity and quality, safe and culturally acceptable, in a political and economically stable environment.

Famine is a severe form of food insecurity and is defined as a 'concentrated decline of food consumption resulting in chronic weight losses for individuals and sharp increases in excess mortality, massive social disruption, and long-term resource depletion' (Webb and Von Braun 1994: 35).

Nutrition security refers to utilization of food (Gross *et al.* 2000: 5). It is a specific form of food security as it prioritizes community's access to essential nutrients, not just calories. Thus, healthy weight as well as obese persons can have a poor nutritional security status, as they might lack vitamins or minerals due to a very monotonous diet based mostly on starch and fat. A poor nutritional status increases the risk of ill-health due to (infectious) diseases. Undernourishment includes acute malnutrition, chronic malnutrition and micronutrient deficiencies (lack of iron, iodine, vitamin A and D being the most important worldwide). Wasting (thinness) and/or nutritional edema refers to acute malnutrition, while stunting (shortness) and/or vitamin and mineral shortage refers to chronic or hidden malnutrition.

Cyclical, chronic and acute food and nutrition security

Food insecurity can differ in nature, depending on its cause. Seasonal or cyclical food insecurity is quite common. A well prepared population has a surplus of food in stock, or is able to buy food elsewhere and can survive this harsh situation. Chronic food insecurity is commonly caused by a lack of assets. Acute food insecurity is caused by a sudden change or shock, such as a drought or a sudden increase of food prices due to civil unrest. The medical signs of malnutrition might help to define whether the causes are acute or chronic; see the section on 'Operationalization and indicators'.

Historical overview of explanations of food and nutrition insecurity

Several theories, based on developments through time, explain the risk of food security in a context larger than the core science of Medicine: it links, for example, to economic, environmental, social or political aspects.

Thomas Malthus (1798) described 'the principle of population' assuming that the human population would grow exponentially and food production could only grow linearly. As the land could no longer support sufficient food, scarcity would be the result. He concluded that demography and biological factors were key, and that the environment would limit human progress. In other words, Malthus's approach focused on the availability of food and the lack thereof as a cause of famine.

After the Second World War, countries with an agricultural surplus disposed of their commodities as food aid. In 1963, the World Food Programme (WFP) was created and the concept of 'food for development' was institutionalized. Food aid was perceived a hinder for self-sufficiency of the receiving countries.

In the 1980s, the green revolution had shown that increased food production was possible. Amartya Sen (1981), an economist, argued in his 'entitlements theory' that famine occurs not only due to a general lack of food in a region, but also due to socio-economic inequalities within that region, such as a lack of purchasing power of specific groups. Distribution of food aid does therefore not automatically lead to less famine and starvation in vulnerable groups, as long as the individual or household cannot escape from a malfunctioning economic or political system (resulting in unemployment, declining wages, rising food prices or poor food-distribution systems).

In the 1990s, the international focus was put on reduction or eradication of hunger and malnutrition. During the same timespan Alex de Waal (1997) showed the damaging effects that war can have on food production and distribution. He gave his view on famine as a process which is highly influenced by political power structures. In other words, famines can be the result of deliberate lack of access and/or availability of food as they can be strategically chosen acts in conflict and war.

Operationalization and indicators

Operationalizing health security

For professionals in humanitarian assistance and protection, reducing excess mortality is the immediate goal. Though existing guidelines provide minimum standards for health services, they do not always address the underlying social and environmental determinants of disease. In this chapter we argue that it is of great significance to understand the root causes in assessing the health context in a humanitarian crisis.

Health security should therefore be analyzed by assessing the health services in terms of access and quality, analyzing the major health threats and their causes, assessing demographic indicators, health expenditure, government and private action and, lastly, by identifying vulnerable groups. This can be done at different levels of analysis (see also Chapter 3):

- internationally and regionally health relates to the presence of relevant medical problems in the area next to the distribution of and access to health care facilities for the affected population;
- nationally it is about equal access to health care facilities and adequate distribution of such facilities over the country;
- at sub-national levels, it concerns the availability of health care services and the possibility for communities and individuals to access these services, including those less visible, as refugees, ethnic minorities, elderly, disabled persons or prisoners.

Health security can be further subdivided into:

- Availability: is there a well-functioning public health system, next to sufficient health care facilities (biomedical and/or traditional, see Note 1), medical staff and medical supplies?
- Accessibility: to what extent are health services accessible to everyone without discrimination?
- Acceptability: to what extent is the health system respectful of medical ethics and is it culturally appropriate, including gender and age?
- Quality: to what extent are the services offered appropriate and of good quality? To what extent are medical personnel sufficiently trained?

Key questions to ask about the health context

On the following pages the most important key questions to be formulated for the particular situation studied are listed.

First of all it is important to measure the number of people affected by health problems caused by the disaster. What is the size of the population? How is their age distribution (infants, children, elderly)?[3] What are the major causes of death (mortality)? What are the major causes of disease (morbidity)? Which minority groups (ethnic, religious, linguistic or sexual orientation) are present? Which health needs do they have? What are the local and individual coping mechanisms?

The second step is to determine the availability, access and quality of health services. Which services are available? Who provides these services (government, non-governmental organizations (NGOs), traditional health, military)? What is the quality of the health services offered? Are there enough qualified health personnel available? Is there any damage to the health infrastructure? Do men, women, children have equal access to health services? Do minorities have the same access to health services as the majority of the population?

Third, it is important to have insight in how health care services are financed and organized. Who had access to the services offered before the disaster? Which groups might have difficulty accessing health services due to lack of financial means? Is there a national safety net? What was the quality of services before the disaster and what changed? Is there a direct link between access to health and political affiliation? Does the disaster situation have a direct effect on access to health care, for example, due to destruction of infrastructure?

Fourth, the environmental context often links directly to the health context. For example, malaria is transmitted by the *Anopheles* mosquitos, which need a humid environment to survive. Change of the environment, for example due to climate change or agricultural development, might lead to the availability of breeding grounds for these mosquitos and therefore to a risk of malaria infections. Another example pertains to cholera which is spread via the bacterium *Vibria cholera.* Transmission to humans mainly occurs via ingestion with contaminated water, when sanitation is of an unsatisfactory level, people do not wash their hands regularly or use dirty water when preparing food. Weak people, such as children or HIV-infected persons, are more susceptible. Thus, key questions to ask are, amongst others, what is the availability, access, quality and utilization of drinking water and sanitation services? Are any preventive programs such as vector control activities active? Are there any relevant environmental topics to be taken into account, such as shelter?

Indicators of the health context

We now explicate possible indicators that can be used to answer the questions posed above. When using the indicators, check whether the units of calculations are converted the same way (per 1,000 or 10,000 persons, for example).

- The crude mortality rate (CMR) is the total number of deaths per year per 1,000 people. The CMR is the most frequently used health indicator to monitor and evaluate the severity of an emergency situation. Take into account that often non-infectious, or so-called non-communicable diseases (NCDs), result in far greater morbidity and mortality than infectious diseases. During an emergency response patients suffering from chronic medical conditions are more vulnerable and at risk when treatment is interrupted, therefore a higher mortality rate can be expected.

- Life expectancy and mortality rates are categorized amongst age and/or gender group, for example Maternal Mortality Rate (MMR), Infant Mortality Rate, Under Five Mortality Rate (U5MR). In the Sphere Handbook one can find the baseline reference mortality data (CMR and U5MR) per region in the world (Sphere Project 2011: 310). The MMR shows the risk of complications during pregnancy and childbirth, which are a leading cause of death and disability among women of reproductive age in developing countries. It is very difficult to measure a reliable MMR as the identification and registration of these deaths, which often occur outside health care facilities, are often estimated. Direct causes are, amongst others, hemorrhage, obstructed labor and abortion. Indirect medical causes are anemia and malaria, nonmedical causes are social status, and lack of access to health care services.
- A low birth weight (percentage of births less than 2,500 grams) indicates a higher risk for infant mortality.
- The case fatality rate (CFR) measures a disease specific mortality rate.
- The number of new cases of a disease that occur during a specified period of time in a population at risk of developing the disease, is called the 'incidence rate' for that disease. The speed at which the incidence rate rises or declines shows whether the situation deteriorates or improves. The height of the rate indicates if the outbreak might end in a potential epidemic or whether the response activities improved the situation. In case the CFR rate for the specific disease is unknown, the CMR emergency threshold can be used (Sphere Project 2011: 310).

  An example of a sudden outbreak of diarrheal diseases with an extremely high CMR was seen in the Rwandan refugees in Goma, Zaire, in July 1994 (Goma Epidemiology Group 1995). Of the 500,000 to 800,000 refugees, almost 50,000 refugees died during the first month after the influx, which means an average crude mortality rate (CMR) of 20–35 per 10,000 per day. In the Sphere Handbook (Sphere Project 2011: 311) the CMR emergency threshold for sub-Saharan Africa is 0.8 deaths per 10,000/day. The situation in Goma was described as 'a public health disaster of major proportions' (Johns Hopkins and Red Cross/Red Crescent 2008: 126, footnote 3), mostly caused by diarrheal diseases. In the second month death rates declined to five to eight per 10,000 per day due to effective interventions based on a rapid analysis of available data.
- The availability and quality of health services can be measured via key indicators such as the health facility utilization rate (the number of outpatient visits per person per year), the average number of total consultations per clinician or nurse per day, the capacity and functional status of existing public health programs (such as an Extended Program on Immunization) or the availability of standardized protocols, essential medicines, supplies and logistics systems.

- Micronutrient deficiencies, when severe enough, lead to physical symptoms (clinical signs), and therefore are part of this list. Anemia is important because iron deficiency is the most prevalent nutritional problem worldwide. It affects an estimated 1.62 billion people (95 percent CI: 1.50–1.74 billion) which corresponds to 24 percent of the population (de Benoist *et al.* 2008). The most commonly used screening methods for the presence of iron deficiency in a population are the measurements of hemoglobin concentration for the presence of anemia (de Benoist *et al.* 2008). Age and stage of pregnancy influence the hemoglobin concentration.
- In many countries worldwide the incidence rate of mental illnesses is unknown or lower than to be expected, due to lack of available data. The number of specialized mental health care personnel is often seriously limited, and traditional services might be the main or only option patients have. In humanitarian situations the incidence of mental illnesses can be expected to be higher due to stress, grief and loss which people are confronted with.

Please note that in cases of weak governance it might be difficult to find reliable data on health security indicators. Often information provided by a variety of sources needs to be compared (so-called 'data triangulation', see Chapter 4).

Some tips as where to search for information

The WHO Global Outbreak Alert and Response Network (GOARN) is a technical collaboration of existing institutions and networks that pools human and technical resources for the rapid identification, confirmation and response to outbreaks of international importance. Institutes as the Pan American Health Organization (PAHO), the Centre for Research on the Epidemiology of Disasters (CRED), the United States Centers for Disease Control and Prevention (CDC), Epicentre-Paris or the Sphere Project offer relevant information based on epidemiological research.

Operationalizing food and nutrition security

Any assessment of the nutritional status of a population needs to be done by qualified personnel. Based on the entitlements theory of Sen, one can extrapolate that only having enough food available is a too limited view on being food secure. Hence, food security is currently specified in the following four dimensions

- availability of food resources;
- access to those resources, also in terms of sufficiency;
- acceptability of food resources, also in terms of diversity and social and cultural appropriateness;
- quality and appropriate utilization in a hygienic and nutritious manner.

Key questions for assessing food and nutrition security

When linking food security to health security it is important to define the major specific health problems related to the lack of certain aspects of nutrition. Look for any clinical signs of malnutrition (stunted, wasted, nutritional edema, vitamin or mineral deficiencies). In case a physical analysis is not possible, one needs to question if such analysis can be made on the basis of a thorough food basket analysis. What kind of nutrition is not covered (energy, fat, proteins, vitamins or minerals)? A nutritious diet is based on a variety of products of good quality.

When concluding there is a food or nutrition security problem, it is important to find out what exactly causes the problem: Is it the availability of food? Or is the access to nutritious food not covered for each person in the target group? How is the utilization of food; do the ones who need nutritious foods, get them? For example, does weaning food contain enough fat to cover the energy intake? And how is the quality of the diet, is it varied enough (ACF International 2010: 36)? What is the cultural context of food consumption? What are the local and individual coping mechanisms (e.g. food sharing patterns)?

Next to the above questions, it is necessary to analyze the economic, environmental and political dimensions of food security. Food needed for consumption is produced, by the consumers themselves, locally, or on international scale, and therefore is an economic activity. The (export) market for cash crops is influencing, and often threatening, the position of local food products, as it limits its agricultural biodiversity. It thus also is important to understand the structure and functioning of local and international markets and its relation to food insecurity.

The intensification of the agricultural system might put a too high pressure on the environment. Monocultures, for example, may exhaust the soil, which makes continuous food production more expensive as fertilizers are needed. The threat of pests and weeds are higher than in a varied ecosystem. A diversity of plant and animal species make the farming system more sustainable as well as resilient to climate conditions.

Linking to political security, how does the political situation have impact on the identified food security problems? Generally, what are the actions taken by governmental and other actors to address food insecurity? For example, what practical results derive from market-based food policy? State involvement in agricultural markets is well known. Examples from Southern Africa show that state involvement has not led to the alleviation of food insecurity (Bird *et al.* 2003). The main reasons are that state interventions are not effectively targeted at the most vulnerable (for example, as was the case regarding the distribution of fertilizer in Zambia), often inefficient (this was the case regarding the distribution of inputs by the Lesotho government) and crowd out the private sector (this happened during the operation of restrictions on maize trade in Zambia, Malawi, Lesotho and Zimbabwe).

One should also question to what extent food is used as a strategic asset for other purposes than eating (i.e. in conflicts, used as a weapon in war)? Conflict often leads to involuntary migration, implying losses of crops or work force. Nevertheless, conflicts do not necessarily lead to a worse food security situation compared to peace time (Keen 2012).

Indicators of food and nutrition security

Food security, just like health security, can also be measured on the individual level, household level and national or regional level. Measurements on the individual level are usually used for assessments of larger groups, to make an overview of the situation and have a starting point for improvements made. Examples of individual level indicators are:

- the Mid-Upper Arm Circumference (MUAC) of children between six and 59 months of age is the main tool to estimate the prevalence of wasting across a population during the first or rapid assessment;
- weight for height and height for age of children under five (and to compare this data to a healthy and well-fed standard population);
- for adults often the Body Mass Index[4] is calculated;
- besides young children, most often pregnant and lactating women are at risk for food related insecurities. Therefore the percentage of pregnant women with iron deficiency anemia as well as the percentage of women exclusively breast-feeding their children up to six months of age are taken into account.

On the household level the following indicators are useful to assess:

- the percentage of income spent on food;
- the energy supply intake (in kilojoules or kilocalories);
- the cost of the basic diet (appropriately defined for each country) as a percentage of family income is an important indicator for access to food.

From an economic perspective the agricultural production of cash crops and food crops can be measured, next to the structure and functioning of markets. Both the World Bank as well as the FAO have developed national food production indexes (respectively http://data.worldbank.org/indicator and http://faostat.fao.org/site/291/default.aspx). They provide time-series and cross-sectional data relating to food and agriculture for some 200 countries.

Some tips as to where to search for information

The Emergency Nutrition Network is an interagency group of humanitarian agencies with the aim to accelerate learning and strengthen institutional memory in the emergency food and nutrition sector. Their website offers a lot of field examples (www.ennonline.net/library).

Some well-known food security assessment tools are the Global Hunger Index (GHI), and the Coping Strategies Index (CSI). The GHI is a statistical tool to measure and monitor hunger in the world by country and by region. The CSI is a tool for rapid measurement of household food security and the impact of food aid programs in humanitarian emergencies.

Aggregation and weighing

Based on the above, we have constructed two tables identifying thresholds for specific health and food security levels (see Tables 9.1 and 9.2).

Health security

Food and nutrition security

Anthropometric tools can be used to measure and quantify the seriousness of food insecurity. The admission criteria for feeding programs are based on these measures.

Measuring in children

Most often the prevalence in the group of children aged between six months and five years is used as reference group for the full population. The measurement of the length or height of a healthy population is spread according to a normal, or Gaussian, distribution. The larger the number of persons in the study group which is relatively low in weight for age (or low in height for age or low in weight for height) compared to the WHO 2006 child growth standards, the more the risk for malnutrition. Find the cut-off points in Table 9.2. Alternatively the MUAC of children between six and 59 months of age is measured in centimeters, and should be performed by trained personnel.

Measuring malnutrition in adults between 20 and 59 years of age

For adults the calculation of the Body Mass Index (BMI) is used (see Note 3). A BMI lower than 18.5 is considered as proof of adult acute malnutrition. Pregnant women and persons with edema should be excluded in the assessment. Elderly are prone to spinal curvature and therefore their height cannot be measured properly. For pregnant women MUAC may be used for screening. They are at greater risk of nutritional insecurity (CDC and WFP 2005: 21). Cut-off points for risk vary per country. For the measurement of nutritional insecurity in a high-weight population, cut-off points for BMI depend on ethnic background.

Table 9.1 Multiple levels of health security

	Level	*Descriptor*
Optimal security	A	*Ideal state of health security*: Effective functioning health services institutions, with trained workforce, giving access to all in need; Control of outbreak of endemic diseases; A working early warning system for outbreak detection; Enough stock and provision of essential medicines; A routine expanded program on immunization service for children aged six months to 15 years, with at least a vaccine coverage of 90%; Access to comprehensive reproductive health information and services, including a communication and transportation system to manage obstetric and new born emergencies; Access for all to safe drinking water, at least 15 liters per person per day, for drinking, cooking and personal hygiene, at maximum 500 meters away from the household and no more queuing time than 30 minutes; Safe and cultural acceptable excreta disposal facilities, cleansing material and sanitary pads; Adequate housing, accessible, safe and affordable; No iodine deficiency.
Suboptimal security	B1	*Average or just below average state of health security*: Functioning state institutions, some worrying indicators in relation to: • prevalence <1% of symptoms of vitamin A deficiency (night blindness); • prevalence of 5–19.9% of visible and palpable goiter in school-age children; • prevalence of 5–20% of non-pregnant women (hemoglobin <12 g/dl) or children 6–59 months (<11 g/dl); • case fatality rates of infectious diseases are below acceptable levels (for example for cholera <1%).
	B2	*Below average state of security*: Functioning health services institutions, but lack of access to those vulnerable (physical, financial, behavioral, cultural, linguistical) and/or lacking trained health workforce; Rise in outbreak of endemic diseases; Housing, which is not well accessible, safe or affordable.

Insecurity	C1	Single indicators or a combination of indicators, which in total threaten to cause damage to a population in the long run: • Prevalence $1 \leq 5\%$ of symptoms of vitamin A deficiency (night blindness); • Prevalence of 20–29.9% of visible and palpable goiter in school-age children; • Prevalence of 20–40% of non-pregnant women (hemoglobin <12 g/dl) or children 6–59 months (<11 g/dl); • Rise in MMR compared to a stable neighboring state with a well-functioning health system.
	C2	Single indicators or a combination of indicators, which in total threaten to cause immediate damage to a population: • A doubling of the regional baseline CMR and/or U5MR requiring an immediate response; • CFRs above the acceptable levels, among others: cholera >1%; *Shigella* dysentery >1%; measles >5%; • A lack of expanded program on immunization services; • No trained health workforce; • Lack of services preventing reproductive health related morbidity and mortality among women, men and adolescents (Minimum Initial Service Package); Lack of water of sufficient quality to be drunk, and/or lack in quantity for cooking and hygienic purposes; Loss of housing.
Acute insecurity	D1	Single indicators or a combination of indicators, which already cause systemic damage to a population health services institutions: • a doubling or more of the regional baseline CMR and/or U5MR requiring an immediate response; • prevalence >5% of symptoms of vitamin A deficiency (night blindness); • prevalence of ≥30% of visible and palpable goiter in school-age children; • Prevalence of >40% of non-pregnant women (hemoglobin <12 g/dl) or children 6–59 months (<11 g/dl); A maximum access to safe water, of 2.5–3 l per person per day, enough for drinking and food under survival needs only.

Source: based on Sphere 2011, CDC and WFP 2005.

Table 9.2 Levels of food and nutrition security, ordered from national/regional to household/individual measurement

	Level	*Descriptor*
Optimal security	A	*Ideal state of food and nutrition security*: • Well-functioning agricultural system; • Households buy food products they prefer; • A food basket available containing an energy content of, at least, 2,100 kilocalories per person per day, of which 10% of the total energy is provided by protein and 7% of the total energy is provided by fat, and an adequate micronutrient intake (average for a group of persons in a large population varying in age, gender, including infants, pregnant and lactating women); • MUAC in children 6–59 months > 12.5 cm; in adults BMI of 22–25 kg/m^2, depending on ethnic group.
Suboptimal security	B1	*Average or just below average state of food and nutrition security*: Functioning state institutions, some worrying indicators: • inflation of currency, rise of agricultural products and processed foods; • households buy food products which are cheaper than preferred; • selling surplus of livestock.
	B2	Below average state of security; seasonal food and nutrition insecurity: • households borrow food, buy food products on credit or consume wild foods, households reduce food portions; • seasonal migration of agricultural workforce.

Insecurity	C1	Single indicators or a combination of indicators, which in total threaten to cause damage to a population in the long run: • selling livestock due to debts; • moderate acute malnutrition in children under 5years weight for height $-3 < $ Z-score < -2 and/or MUAC $11.5 \leq 12.5$ cm; in adults BMI of 18.5–22 kg/m^2; • rise of number of people living with HIV or TB patients compared to earlier in time.
	C2	Single indicators or a combination of indicators, which in total threaten to cause immediate damage to a population: • droughts or floods influencing the agricultural system; • migration of agricultural workforce; • damage to infrastructure which makes accessing the food market difficult, closure of warehouses and agro-processing facilities; • households beg for food, buy food products on credit or consume their seed stocks; some members of the household skip one or two meals per day; • low birth weight as a measurement of mother and infant nutritional status; in adults BMI of 18.5 kg/m^2.
Acute insecurity	D1	Single indicators or a combination of indicators, which already cause systemic damage to a population: • households skip meals for whole days; • physical appearance of vitamin and mineral deficiencies (see health security descriptors); • severe acute malnutrition in children under 5 years weight for height Z-score < -3 and/or MUAC < 11.5 cm and/or nutritional edema in children; • pregnant women at risk in emergencies when MUAC < 21 cm; • in adults BMI < 18.5 kg/m^2.

Source: based on CSI 2008, Famine Early Warning System (www.fews.net), CDC and WFP 2005.

Notes

1 The Biomedical Model of Health was developed over the last two centuries in Western society. It is a model that regards disease and ill-health as the consequence of certain malfunctions of the human body. The model is based on the rational application of technical scientific medicine (Bilton *et al.* 2002: 356).

2 The term 'sanitation' refers to excreta disposal, vector control, solid waste disposal and drainage.

3 Use UN agreed cut-off points for age.

4 $\text{BMI} = \frac{\text{mass (kg)}}{(\text{height (mm)})^2}$

A BMI <18.5 is considered underweight; the cut-off point for overweight and obesity vary per global region (from 23 on), as genetic differences between ethnic groups appear to play a role in the risk for health problems.

10 The environmental context

Peter D.M. Weesie

The environment: definitions, views and facts

Current definitions of the notion 'environment' include 'the totality of the physical conditions on the Earth or a part of it, especially as affected by human activity' (*The Concise Oxford Dictionary* 1998: 452), 'the external conditions or surroundings, especially those in which people live or work' (*Collins English Dictionary* 1994: 520) and 'the circumstances, objects, or conditions by which one is surrounded' (Merriam-Webster Online Dictionary). All these definitions characterize the environment as being connected to humans, as if without humans, or before their existence on Earth, there would be no environment. This anthropocentric view is exemplified, and supplied with a humanitarian perspective, by the Sphere Handbook (2011: 14) that understands the environment 'as the physical, chemical and biological elements and processes that affect disaster-affected and local populations'.

In contrast, from a natural sciences' point of view, the environment is simply seen as 'all living organisms and non-living matter on Earth'. It came into existence with the Earth itself some 4.5 billion years before present, and modern humans became part of it from their appearance around 200,000 years ago. It is composed of four basic building blocks that are the lithosphere (the Earth's crust, rocks and soils and underlying tectonic plates), the hydrosphere (all salt and fresh waters), the atmosphere (gasses up to an altitude of 100 km) and the biosphere (all living beings). The four spheres are closely interlinked by large exchanges of matter and energy called biogeochemical cycles because of their cyclic nature and the numerous biological, geological and chemical processes that are involved. An example is the water cycle: the continuous transport of massive amounts of water from the seas through the air to the land (where the rains are the sole source of fresh water) and back to the seas. Other important cycles among the four spheres include the carbon cycle, oxygen cycle and phosphorus cycle.

The environment is composed of ecosystems…

The common ground in the above definitions is that they all consider the environment as being composed of living beings and non-living elements, analogously to the distinction between the biosphere and the other three spheres mentioned above.

The living beings include all humans, animals, plants and other life forms (such as bacteria that are neither plant nor animal) on Earth. All these are in close interaction with one another, and with elements from the other three spheres. These interactions take place in a structured way within ecosystems. An example is the human body that exchanges intensively with the atmosphere (e.g. by inhaling oxygen and exhaling carbon dioxide), with the hydrosphere (e.g. by drinking water) and the biosphere (e.g. by eating plants, fruits and animals, and by harboring about 100 trillion microorganisms in the intestines). In this way humans, and all other living beings, are part of an ecosystem that comprises all living beings and dead matter with whom they are connected through uncountable exchanges of matter and energy. Many types of ecosystems exist, with and without humans, terrestrial and aquatic, highly complex with many species (e.g. tropical forests), and less complex with fewer species (e.g. deserts), and on scales varying from micro (e.g. the intestinal flora) to macro (e.g. a forest or lake ecosystem). They all share the following characteristics:

- they are composed of (living and dead) parts with many interactions among them in the form of structured processes, such as feeding on plants or on animals within so-called food chains;
- these ecosystem processes consist of exchanges of matter and energy and confer a high degree of organization to the ecosystems;
- ecosystem processes balance each other out in a dynamic way, if not within the ecosystem itself, then with neighboring ecosystems, resulting in a permanent state of so-called dynamic equilibrium;
- ecosystem processes are partly feedback mechanisms, whereby some proportion of the output is returned to the input, leading to positive feedback where original change is amplified, or negative feedback where change is counteracted;
- because of their dynamic states of balance, ecosystems have a capacity to recover by themselves from partly damage or destruction (called resilience);
- there are limits to resilience, beyond a certain level of damage an 'ecological threshold' can be passed or a feedback loop disturbed, after which the ecosystem cannot come back to its original state anymore. For example, there are many cases of permanent deforestation around refugee camps within tropical forest and savanna ecosystems (e.g. as described by Beaudou and Cambrézi 1999: 7);
- by processes playing out, ecosystems can be seen as functioning systems that maintain themselves and other ecosystems by delivering so-called ecosystem services (see below) to their constituent parts, such as the production of plant biomass and oxygen, the uptake of carbon dioxide and many more.

... and all ecosystems together make up the Global Ecosystem/the environment

All ecosystems are interlinked through the processes mentioned above, and all together they make up the 'Global Ecosystem', also called 'Earth system'.

Following from the above, we consider the notion Global Ecosystem to be synonymous with 'environment'.

This system view has developed over the last decades; it sees the Global Ecosystem/the environment as a worldwide, continuously functioning biophysical system, composed of the four spheres and organized in countless local- and regional-scale ecosystems that are all functioning, interacting and contributing to the functioning of the whole. The Global Ecosystem possesses all ecosystem traits, including the running of resilient ecosystem processes and delivery of ecosystem services, and the existence of feedback loops and thresholds. All this happens on different scales, from micro to macro, from local to regional and from continental to global.

Ecosystem services and human well-being

Ecosystem services have been defined as 'the benefits people obtain from ecosystems' (Millennium Ecosystem Assessment 2003: 53). Again, this is an anthropocentric definition, from a natural sciences' perspective ecosystem services are rather seen as 'the favorable living conditions provided by ecosystems to living beings'. Indeed, all living beings for their physical survival depend on the materials and substances provided by ecosystems, such as clean air, water, food and shelter. Other ecosystem services that are important for human societies include climate and water regulation, the provision of fuels and building materials, and many kinds of cultural services. A comprehensive overview of ecosystem services is presented in Figure 10.1 that gives an idea of the many ways in which human well-being depends on the use of ecosystem services.

Since ecosystems exist on all scales from very small to very large, the scale on which ecosystem services are delivered also differs widely. Each individual green plant produces some oxygen which can be seen as a modest ecosystem service, but all green plants together maintain a sufficient level of oxygen in the atmosphere for animals and man to live on. Climate and water regulation are global processes. Thresholds on these scales are called tipping points and if they are passed, the stability of the entire system is disrupted which may adversely affect large areas and their populations (see below).

Environmental resources

So, we understand the Global Ecosystem/the environment as a worldwide, functioning system composed of interrelated ecosystems, delivering services to their own living beings and to those of other ecosystems, with the only external input being the radiation from the sun. The ecosystem components, processes and services can be considered as environmental resources that biologically, chemically and physically underpin life on Earth. Indeed, the environment constitutes the physical basis of human well-being in all its dimensions: physical, mental, social and cultural. Flourishing human societies are built on 'healthy' environmental

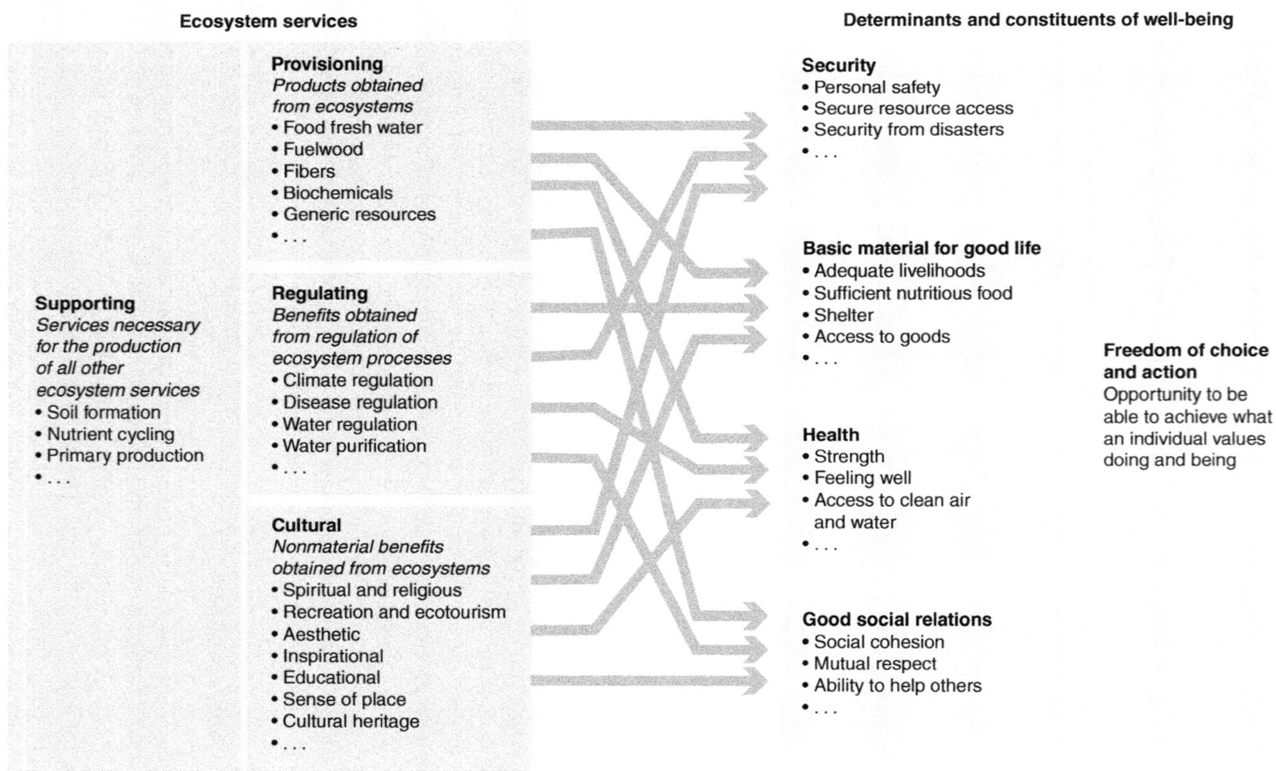

Figure 10.1 The many ways in which human well-being depends on ecosystem services (source: after Millennium Ecosystem Assessment 2005).

resources, as shown by the many great civilizations from the past that went down and ultimately perished partly as a consequence of their poor environmental management practices (Diamond 2005).

Environmental security

In line with the above, environmental security can be understood as the threat to human well-being posed by declining environmental conditions. In Maltais *et al.*:

> environmental security is understood as the notion that some forms of environmental change in combination with socio-economic drivers increasingly threaten basic human interests and thus should be considered high-order threats to human well-being in the same way a military threat is a high-order threat to human well-being.
>
> (2003: 4)

Human impact upon the environment

Humans are a natural part of the environment, but with their growing numbers and technological means have an increasing impact on its composition and functioning, and hence on the ecosystem services provided. Positive impact may result from careful land and urban planning (that is, in a way that local ecosystems can continue to function), and from nature rehabilitation and conservation projects. More often, human impact leads to environmental degradation, that is, the deterioration of environmental resources in terms of both quantity and quality. Environmental degradation can have two opposite causes: depletion of environmental resources (and thus of the environment), and pollution. Hence a division of environmental degradation into depletion and pollution problems is warranted.

Depletion of resources

Depletion of environmental resources leads to their absence and takes place in many ways. Well-known depletion problems include the following:

- Deforestation: the rate of deforestation is decreasing, but is still alarmingly high. Around 13 million hectares of forest were converted to other uses or lost due to natural causes each year in the period 2000–2010 compared with 16 million hectares per year in the 1990s (FAO 2010: XIII). This is equivalent to a loss of more than 300 km^2 per day.
- Overfishing: currently 11 of 15 major oceanic fisheries are in decline due to heavy fishing (Chiras 2013: 189).
- Over-pumping: half of the world's population live in countries where water tables are falling as aquifers are being depleted with the help of powerful pumps, mostly for irrigation (Brown 2011: 22–3).

For example, in about 20 countries, including China, India and the United States – that together produce half the world's grain – water tables are falling and wells are running dry. And since 70 percent of world water use is for irrigation, water shortages can quickly translate into food shortages (Brown 2011: 22–3). Thus over-pumping, and the ensuing insecurity of water provision, may affect social and food security.

Pollution of resources

Pollution of environmental resources leads to their unusableness. It consists of the contamination of air, water, land and living beings mostly by man-made waste in many forms: solid, liquid, gaseous and chemical. Well-known pollution problems include air pollution leading to acid rain and global warming, oil spills leading to land and water pollution, and discharge of solid and chemical waste, and of sewage in water and on land. Of these, air pollution with carbon dioxide is one of the worst problems, leading to climate change and global warming. The industrialization of western countries since the 1850s was possible thanks to the large-scale combustion of fossil fuels such as coal and oil; enormous quantities of carbon dioxide have been released in the atmosphere since then. Carbon dioxide is heat absorbing and – in very small quantities – a natural component of the air, but the current levels of carbon dioxide concentration in the air exceed those existing before 1850, by approximately one-third. According to the IPCC (2013: 17):

> Human influence has been detected in warming of the atmosphere and the ocean, in changes in the global water cycle, in reductions in snow and ice, in global mean sea level rise, and in changes in some climate extremes. This evidence for human influence has grown since AR4 [the IPCC assessment report of 2007].

And, 'it is extremely likely that human influence has been the dominant cause of the observed warming since the mid-20th century'.

Warming of the atmosphere and the oceans means that they contain more energy, which translates into more-frequent, more-powerful and more-destructive tropical storms. This may be illustrated by a record '10' typhoon in Japan in 2004; the 2005 Atlantic hurricane season that was the worst seen so far (Brown 2011: 75–6) with Wilma, Rita and Katrina among the six most intense hurricanes ever recorded; and Hurricane Sandy in 2012 that turned out to be the largest (as measured by diameter) Atlantic tropical storm on record (WBZ-TV 2012). Many other examples of extreme weather events in recent years can be given. Although it is not possible to prove scientifically that these events individually or collectively are linked to climate change, the likelihood that they do is high.

Atmospheric temperature increase is highest in the (Sub-)Arctic, where the continuing melting of the Greenland Ice Sheet and loss of sea-ice are considered to represent climate tipping elements: areas where an environmental threshold or tipping point is passed and the state of balance disrupted, thereby causing unstoppable climate change (see Lenton *et al.* 2008, Duarte *et al.* 2012). Indeed, the feedback loops that exist in these areas and that confer a certain stability to the climate system may be disturbed. For example, with the disappearing ice sheets the reflection of the sun's radiation back into space will also decrease, leading to further warming of the sea water (positive feedback). Moreover, there is scientific consensus about the fact that if the Greenland Ice Sheet were to melt entirely, it would lead to a global sea level rise of seven meters or so.

Combinations of depletion and pollution

In recent years, a new kind of environmental hazard emerges that results from a combination of depletion and pollution of environmental resources. For example, the 2010 Pakistan floods in July and August led to nearly 2,000 deaths, more than 20 million people were affected, mostly by destruction of property, livelihood and infrastructure, over a million livestock drowned, and some two million homes were damaged or destroyed (Brown 2011: 4). The direct cause of the floods were the unusually heavy monsoon rains in the mountains of northern Pakistan in late July, possibly related to global warming. But there are other factors that appear to have contributed to the flooding (Brown 2011: 5):

- record high temperatures, possibly related to global warming, were registered in the western Himalayas in the preceding April and May and led to the Indus river unusually swollen with melting water;
- heavy deforestation has taken place in the Indus river basin since the 1960s, seriously reducing the water retention capacity of the area;
- high pressure of livestock (an estimated 149 million cattle, sheep, goats and water buffalo) that strips the country of the remaining vegetation, increasing run-off.

In this case, different pollution and depletion problems converged to result in what turned out to be a devastating disaster with about one-fifth of Pakistan's total land area under water, and an enormous impact on human well-being.

Resuming, degradation of the environment – through any form of pollution or depletion of environmental resources or a combination of these – directly and indirectly affects the functioning of the Global Ecosystem/the environment and the delivery of ecosystem services, thus undermining human well-being. Indeed, environmental degradation often has serious effects on health and on the availability of food, housing, infrastructure and natural resources, coupled with social effects on availability of education, work, health care, social services and safety, and on economic and political conditions. These effects are manifested on all spatial and time scales, and at the individual, family, community and country levels.

Environmental and natural disasters

Human impact on the environment can go so far as to create environmental disasters, whereby widespread destruction and distress is caused. An example is the 1986 Chernobyl disaster in Ukraine with large quantities of radioactive particles released into the atmosphere, causing thousands of cancer deaths and cases of radiation related disorder, and widespread contamination of the environment (soils, buildings, surface- and groundwater, flora and fauna). Oil spills are other well-known environmental disasters caused by man.

Environmental disasters with an entirely natural origin, that is, without apparent human interference, are often called natural disasters. They include volcanic eruptions, most earthquakes and tsunamis (caused by landslides, undersea earthquakes and volcanic activity) and meteor impacts. They may occur on all time scales varying from seasons and years to decades and even millions of years. There is fossil evidence of volcano outbursts and big events of drastic climate change of many millions of years ago. They can be seen as normal phenomena of the functioning Global Ecosystem that represent temporary intensifications of the cycling of matter and energy among the four spheres (see above).

The origin of an environmental disaster may thus be entirely natural, but it can also be entirely man-made. In between these extremes there is a growing number of environmental disasters that have a natural origin, but of which the impact on human populations and their environment is aggravated by the effects of human action. The 2010 Pakistan floods have been described as the largest natural disaster in Pakistan's history (Brown 2011: 4). However, as is shown above, the underlying causes are at least partly related to human activity, so the term 'natural' is not quite appropriate here.

Also other climate related events such as droughts, fires, tropical cyclones (called hurricanes in the Americas and typhoons in Asia), tornadoes and other strong wind storms may have a natural origin but as pointed out above, can be – and often are – exacerbated or even caused by the effects of human action. Disasters like floods, mudslides or avalanches may be caused, or exacerbated, by deforestation in hilly regions or channelizing streambeds. Water regulation is an important ecosystem service of forests: they absorb excessive rainwater as a sponge and release it bit by bit. If natural forests are replaced by agricultural or cattle grazing areas (in itself the main cause of deforestation worldwide) the sponge function disappears to some extent, thus rendering the downstream regions and countries more vulnerable to flooding and, possibly, disaster.

Environmental disasters and vulnerability

Humans have always had to deal with environmental disasters (see Table 10.1). Most of them are beyond human control. However, we have learned to better predict their occurrences and mitigate their effects. Taking precautions and timely evacuation of people may reduce the impact of the event on the human population. For example, before the tropical cyclone Phailin made its landfall on

Table 10.1 Major environmental disasters and their impact

Estimated deaths caused by major environmental disasters, 1960 to 2000			
Disaster type	*Estimated total deaths*	*Major events*	*Deaths*
Tropical cyclones	850,000	Bangladesh 1970	500,000
Earthquakes	650,000	China 1976	250,000
Floods	60,000	Vietnam 1964	8,000
Avalanches, mudslides	50,000	Peru 1970	25,000
Volcanic eruptions	36,000	Colombia 1985	23,000
Mortality from major environmental disasters, 2008			
Disaster type	*Date*	*Country*	*Deaths*
Tropical cyclone	May 2, 2008	Myanmar	138,366
Earthquake	May 12, 2008	China	87,476

Sources: after McKinney *et al.* (2013: 149) and Wright and Boorse (2011: 431).

the east coast of India in October 2013, the government had evacuated some 800,000 people living along the coast, thus limiting the number of deaths to 23. In the same state of Odisha, a cyclone in 1999 killed more than 10,000 people. This time the government was warned by the meteorological department that, thanks to India's space program and weather satellites, was able to predict the course and intensity of the cyclone (NRC 2013: 10–11).

Taking successful precautionary measures depends on the presence of functioning warning systems and adequate governance at the local, provincial and national levels. These often do not exist, especially in developing countries. Therefore, these countries are less well prepared for environmental disasters than other countries, and their populations are more vulnerable to the impact of them. The magnitude of the impact on the population may widely vary, according to the extent and intensity of the environmental event, population density in the area affected and location. For example, vulnerability is high in areas such as coastal lowlands, river deltas and floodplains that are prone to flooding and which are, unfortunately, often densely populated.

Environmental refugees

The term 'environmental refugee' was introduced in the 1970s by Lester Brown of the Worldwatch Institute. It became common usage after a 1985 United Nations Environment Programme policy paper entitled 'Environmental Refugees' by Essam El-Hinnawi (Lehman 2009). However, the terminology surrounding the term 'environmental refugees' still is under academic debate, some arguing that a firm definition cannot be given (see Black 2001) and UNHCR using the term 'environmentally displaced persons' instead (Christiansen 2010: 9). However, Brown (2011:

72–83) gives many examples of the following five categories of environmental refugees, all of whom originate from environmentally induced migration, such as:

- those fleeing rising sea levels, especially from densely populated low-lying coastal cities, river deltas and island countries;
- those permanently fleeing destroyed land after the passing of destructive storms;
- people who flee from advancing deserts;
- people who are forced to move from places where water tables are falling;
- those who try to escape toxic waste or dangerous radiation levels (e.g. more than 350,000 people were forced to resettle after the Chernobyl disaster described above).

Disaster-struck and refugee populations, humanitarian assistance and the environment

If after a disaster humanitarian assistance comes into action, a threefold interaction with the disaster-struck or refugee population and the environment will be established (Figure 10.2). The environment may (continue to) exert a negative impact upon the population, as well as on the humanitarian activities (the black arrows pointing from the environment to the population and humanitarian aid), for example through the presence of chemical pollutants, debris and carcasses that contaminate land and water, bad weather conditions and (new) hazards such as floods, mud and landslides (see UNEP 2007, 2008: 2–4).

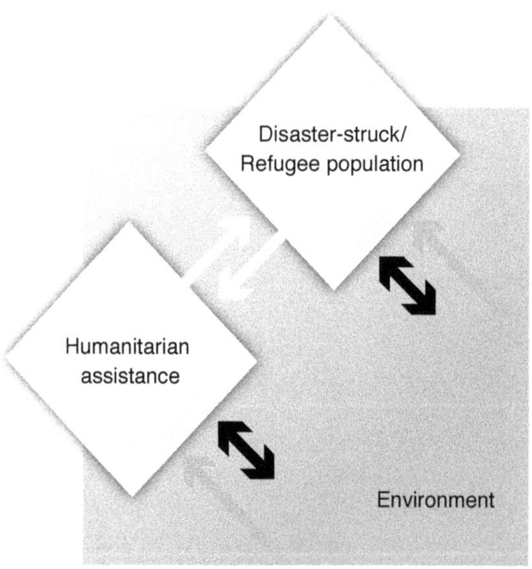

Figure 10.2 Interplay between disaster-struck/refugee populations, humanitarian assistance and the environment.

At the same time, the environment plays an essential supportive role both for humanitarian actors and the struck population (grey arrows from environment), while either of them often use the environment in an unsustainable way and/or have a negative impact upon it (black arrows to environment). The interplay between disaster-struck/refugee populations and humanitarian assistance is not considered here (blank arrows).

Disaster-struck or refugee populations in the immediate aftermath of a disaster or upon arrival in a new location, (continue to) depend for their survival on environmental resources in a number of ways, such as the collection of (fresh) wood for building material and fuel, tapping of water from surface and/or subsurface waters and the gathering of all kind of food items. Environmental damage is often the drawback of such activities. There are many examples of serious environmental degradation in the wide surroundings of refugee camps, through land clearing for settlement, disposal of human, medical and solid wastes, forest degradation and complete deforestation leading to soil erosion, depletion and/or contamination of water resources, and the loss of genetic resources and of wildlife through poaching (e.g. Biswas and Tortajada Quiroz 1996, UNHCR 1996: 1, Sanders 1997, UNHCR 1997). This may not only cause deterioration of refugees' welfare but may also lead to competition with local communities over scarce resources. Also, concerns for the environment may influence a country's decision to provide asylum to refugees (UNHCR 1998).

The humanitarian sector also often depends on locally available environmental resources for its operations, such as groundwater and trees, for provisioning the population in distress and constructing shelters. By doing so, humanitarian activities can cause new environmental impacts (UNEP 2008: 4). In UNEP 2007 and 2013, many examples are given of unsustainable practices of the humanitarian sector in the past, including improper management of solid, human, chemical and health care waste, over-pumping of groundwater, pollution of surface and groundwater, deforestation, soil erosion, inappropriate site selection, etc. However, by identifying humanitarian activities that are designed to minimize negative environmental impact in the 2013 joint UNEP/OCHA info sheet (UNEP 2013), the dedication to reduce adverse environmental effects is shown.

Humanitarian action and environmental management

In 1987, the first UN provision ever in the field of environment and refugees was taken (UNHCR 1990). The first edition of UNHCR's *Environmental Guidelines* for assessing and dealing with environmental impacts in refugee situations appeared in 1996, followed by a revision in 2005. Herein, it was stated that 'environmental concerns are still not always dealt with in a consistent manner in refugee and returnee situations' (UNHCR 2005: 5), notwithstanding the production of a series of handbooks and other publications on environmental management in refugee and returnee operations (for an overview see UNHCR 2009a).

The humanitarian sector at large took up the issue in the early 2000s, with the most important contributions, in addition to the continued efforts of UNHCR, being the series of disaster management working papers of the Aon Benfield UCL Hazard Centre, the many publications by Charles Kelly (e.g. 2001, 2004, 2005a, 2005b, 2008), and the country series of environmental assessments by UNEP.

Kelly (2013) presents a synopsis of key organizations working at the integration of environmental issues into humanitarian assistance, of policy and guidance to this end, and of the tools used. Despite the multitude of organizations, policies and methods, and although the importance of mainstreaming the environment in humanitarian response has been stressed in various publications (e.g. Kelly 2004, ERM 2007) and is addressed by, for example, the web-based resource center 'Mainstreaming the environment into humanitarian action' (UNEP 2011), Kelly (2013: IV) recognizes three factors to which the gap in addressing environmental issues in humanitarian interventions is due:

- different understandings of the term 'environment';
- a weak or poorly documented evidence base to justify why paying attention to the environment is critical;
- defining environment as a cross-cutting issue[1] reduces immediate relevancy to humanitarian interventions.

What does not help in this respect, is that only few people working in the humanitarian sector have a background in natural sciences.

This chapter attempts to contribute to filling the observed gap by addressing the points mentioned above. In the following sections we will discuss some of the tools, such as environmental indicators, and methods, such as environmental impact assessment, that can be used to incorporate environmental issues in humanitarian interventions, with reference to their place in the humanitarian project cycle.[2] Many other tools, methods and guidelines are given by Kelly (2013) and are available through the environment section of the humanitarian response portal at www.humanitarianresponse.info/topics/environment (accessed April 12, 2014).

Environmental indicators

Environmental indicators inform about the state of the environment at a given location. In the natural sciences they are often numerical values, based on quantitative measures or statistics of selected environmental characteristics that are tracked over time, at scales that may vary from local to continental. They may be drawn from any environmental compartment (i.e. air, water, land/soil, living organisms, corresponding to the four spheres mentioned above), and combined into environmental indices. For example, to assess water quality there are many parameters at hand, chemical (e.g. nitrate content), physical (e.g. temperature) and biological (e.g. presence of microbes), that can be used individually or in combination. Because the number of measurements that can be taken is enormous, in environmental management often those indicators are used that are relevant for human health, or for the health of other living beings. For example, an important quality parameter of drinking water is the number of fecal bacteria *Escherichia coli*[3] per water volume unit.

In humanitarian assistance, also 'environment related' indicators are used, that can have a biological or ecological character (e.g. 'change in vegetation cover' and 'area of degraded land') or a more socio-economic or juridical nature (e.g. 'energy saving cooking habits practiced' and 'refugees have usage rights to natural resources'), or be related to human health (e.g. 'safe water quality'). All these examples are drawn from Module V of UNHCR's FRAME toolkit (see below).

Environmental indicators can be used individually or in combination to inform about the quality of the environmental compartment they are drawn from. For example, many countries and international bodies have set water quality standards, e.g. the WHO guidelines for drinking-water quality. By comparing gauged indicator values to the norms set, an objective judgment about the quality of the measured water can be given. However, it is not within the scientific method to combine different indicator – or index – values, for example for water and for air, so as to arrive to a single figure – or place on a scale – indicating overall environmental quality. The same applies to environment related indicators that also have inherently different characteristics which cannot be averaged out to a single value, let alone in combination with environmental indicators.

However, environmental and environment related indicators allow for objectively monitoring of the state of the environment, by comparing indicator or index values measured at different times. Therefore, they are important tools to account for the environment in all phases of the humanitarian project cycle, as applied by some of the methods described below. In the initial assessment phase they tell something about the environmental context in which an intervention is to take place. In the design phase, they may help to formulate environmental goals and make part of environmental (impact) assessment methods that are to be included in the project plan. During the implementation phase they are the key tools for environmental monitoring, and finally during the evaluation phase they inform on the achievement of the environmental goals set in the beginning.

Environmental assessment

Basically, an 'environmental assessment' (also called 'environmental context assessment') is an appraisal of the state of the environment, based on the information given by a number of environmental and/or environment related indicators. In practice, the term is often confounded with 'environmental impact assessment', which is meant to assess the impact a project or intervention may have upon the environment (see below).

Environmental (context) assessment may serve as an *ex ante* baseline assessment during the initial assessment phase of a humanitarian project. However, due to obvious time and logistical constraints in the situation of sudden-onset disasters or arrival of refugees, this may be difficult to apply in practice. Perhaps it is for that reason that environmental (context) assessment does not yet seem to take place in a systematic way in the initial phase of humanitarian action, although parts of the following structured methods may allow to do so: Emergency Shelter Environmental Checklist (Kelly 2005b), Environmental Needs Assessment (UNEP 2008), Camp Management Toolkit (NRC 2008), Post-Disaster Needs Assessment (UNDP 2009) and Green Recovery and Reconstruction Toolkit (WWF and ARC 2010). This list is not exhaustive, and also REA (see below) may be used to this end.

Nevertheless, because of the great importance of having basic environmental information in the initial phase of humanitarian action, it would be good practice to ask common sense questions about issues such as the size of the involved population, the kind of landscape (desert, savannah, forest,…), the extent to which environmental resources are available (surface water, depth of groundwater, wood,…), the presence of protected or ecologically sensitive areas in the vicinity, presence or absence of effective environmental governance (legislation, government bodies), accessibility of the area and other relevant questions (after Groupe URD and Joint UNEP/OCHA Environment Unit 2013).

Environmental impact assessment

Environmental impact assessment (EIA) is an exercise that identifies the potential positive and negative environmental impacts of an anticipated intervention or project and the appropriate mitigation measures, while accounting for the relevant environmental, social and economic aspects. Hereto, use is made of environmental and socio-economic indicators that are aggregated, weighed and scaled to inform the assessment. Under normal conditions, EIA is mandatory, proactive, comprehensive and applied before deciding on major infrastructural projects such as the construction of buildings, dams or roads. Obviously, in the case of a disaster or refugee situation the context is quite different and the standard EIA process has been adapted so as to be applicable in case of a sudden onset event at an unpredictable location, to be reactive and possibly partial in coverage, to have heterogeneous and dynamic beneficiaries, to be based on readily available data, to be completed in hours to weeks and to allow direct 'life saving' activities to have priority above environmental goals, while the cost depends on available funding (see Kelly 2004, 2005a: 1).

Rapid Environmental Impact Assessment in Disasters

The Guidelines for Rapid Environmental Impact Assessment in Disasters (REA) were developed by CARE International and the Aon Benfield UCL Hazard Centre (Kelly 2013: 22). They are based on the concept that identifying and incorporating environmental issues into the early stages of a disaster response will make relief activities more effective and lay a foundation for a more comprehensive and speedy rehabilitation and recovery (Kelly 2005a: 2). The REA identifies, frames and prioritizes environmental issues in such ways as to allow the negative impacts to be minimized or avoided during the immediate response to a disaster. It does not provide answers as to how resolve the critical issues identified, but gives sufficient information to take appropriate action (Kelly 2005a: 3). REA is to be applied in the initial and design phases of the humanitarian project cycle.

Environmental assessment in refugee and returnee situations

The Framework for Assessing, Monitoring and Evaluating the Environment in Refugee-related Operations (FRAME) is a set of six modules designed for assessing, addressing and evaluating the impact of refugees on the environment in a participative approach (see UNHCR 2009b, Kelly 2013: 24). The FRAME toolkit makes use of a list of 50 environment related indicators, features modules on (rapid) environmental assessment, community environmental action planning (CEAP)[4] and evaluation, and pays attention to planning and monitoring. Thus all phases of the humanitarian project cycle are concerned. The different components can be used individually or in combination, also with other tools, e.g. CEAP and REA are recommended in 'D projects' of the 'Environment marker' method described below.

Dealing with indicators within FRAME

The many methods of environmental assessment and evaluation referred to in this chapter use different manners of dealing with the used indicators. It would be beyond the scope of the chapter to treat them all; however, to give the reader an idea we will briefly describe the way that FRAME's Module V makes use of indicators (see UNHCR 2009b).

As explicated above, the module uses a list of 50 environment related indicators, that 'have been developed adhering to the four principles outlined in UNHCR's Environmental Guidelines [see UNHCR 2005] – an integrated approach; prevention before cure; local participation in decision-making and actions taken; and cost-effectiveness'. Which indicators to use, depends on 'thorough problem analysis and definition of programme and project objectives', and is 'based on cultural, environmental, and/or agro-ecological contexts'. Then for each selected indicator a standard 'score card' is filled in by selecting a score that measures the indicator's performance. Two types of score cards can be used. Quantitative score cards assign a score as a percentage of the target reached. For example, if in a reforestation project an indicator is the number of seedlings produced versus a target number, a score from 1 to 5 may be used to indicate the percentage of seedlings produced – representing the indicator's performance. Qualitative score cards are measured through a series of yes/no questions in which each answer is assigned a score. For example, a qualitative score card measuring erosion around camps may feature the yes/no question whether areas prone to soil erosion have been identified. Then, at the next stage, the employed score cards are collected and the gauged indicators are analyzed, resulting in an insight into the state of the environment. Filling in the score cards at regular time intervals (e.g. every three or six months) then allows for identifying trends in project results and monitoring of positive and negative impacts of the project on the environment and beneficiaries' welfare (after UNHCR 2009b).

Environmental assessment focusing on hazardous chemicals

The Flash Environmental Assessment tool (FEAT) has been developed at the request of OCHA by the National Institute for Public Health and the Environment (RIVM) of the Netherlands (RIVM 2009). The tool helps to identify existing or potential acute environmental impacts from released hazardous chemicals that pose risks for humans, human life-support functions and ecosystems, following sudden-onset natural disasters. FEAT is typically used in the initial, design and assessment phases of the humanitarian project cycle.

Environment marker

In 2011, UNEP developed the CAME approach[5] to reduce the environmental impact of UN humanitarian interventions (see UNEP 2012, 2014a). Here, the following four steps are to be taken: contextualize new projects while accounting for environmental vulnerabilities in the project areas; assess new projects for potential negative environmental impacts; mitigate those impacts by modifying the project design or compensate for them; enhance environmental benefits. The assessment phase uses an 'Environment marker' (see UNEP 2014c) to classify projects according to their potential positive or negative impact on the environment. The environment marker is accompanied by country specific 'tip sheets' (see UNEP 2014b) that provide recommendations per sector on how to better integrate environment in envisaged humanitarian activities. The CAME strategy is meant to integrate the environment throughout the entire humanitarian project cycle.

Conclusion

More than 25 years after the UN's first official recognition of the detrimental environmental impact of refugee camps (see UNHCR 1990), and after many crises since that time, including the environment in humanitarian action is still not common. However, many initiatives from within the sector are now being taken to increase environmental awareness among the personnel and to 'green' the humanitarian project cycle. Important steps taken here include the systematic appointment of Environmental Field Advisors in UN relief and recovery operations, the training organized by the Joint UNEP/OCHA Environment Unit (JEU) and partners, and the online learning modules offered by the JEU through the Environmental Emergencies Centre (EEC) at www.eecentre.org.

Notes

1 The 2005 Humanitarian Reform Agenda defined 'environment' as a cross-cutting issue in the humanitarian Cluster Approach.
2 In this book we understand the humanitarian project cycle as being composed of the following phases: initial assessment (identifying), design (formulating and planning), implementation and monitoring, evaluation (see Groupe URD and Joint UNEP/OCHA Environment Unit 2013).
3 *Escherichia coli* is a common bacterium in human and animal feces.
4 Community environmental action planning (CEAP) is designed to ensure participation of the local population.
5 So far, implemented in Afghanistan, Sudan and South Sudan.

11 From context analysis to intervention design

Liesbet Heyse

In Chapters 3 to 10, we have outlined how to make a thorough analysis of the context of a humanitarian crisis. Various contexts for analysis have been identified – such as the health, political and social context – and for each specific context guidelines have been given as how to investigate the nature and characteristics of this specific context. In addition, the radar graph has been introduced as a tool to compile and compare evidence collected for various context dimensions. The next question of importance to humanitarian organizations and aid workers is how to decide – given a certain humanitarian crisis context – if humanitarian aid is required and, if so, how to arrive at a proposal for a humanitarian aid project or program on the basis of the information collected in the context analysis. In other words, how to adopt an evidence-based approach to humanitarian programming? This question is addressed in the next chapters of this book.

A stepwise approach to evidence-based humanitarian programming

In this chapter, we will develop an evidence-based approach to programming. As elaborated previously, the context analysis gives information about the core humanitarian problems and vulnerabilities in the context, next to a clear overview of existing capacities in the crisis context. Once these vulnerabilities and capacities have been identified, it is possible to identify areas in which intervention is needed (for example in terms of food, shelter, water or psychological care), where capacities are low and where vulnerability is high. The next question is how to use this information to arrive at a programming decision.

We answer this question by proceeding with the step of Intervention Analysis. We do this in two intermediate steps in this chapter: first, we identify a set of *thinkable* interventions, which is then reduced by means of the method of theory-based *ex ante* evaluation to a set of *suitable* interventions that fits the humanitarian problems and context. In the chapter on stakeholder analysis (Chapter 12) a final step is outlined that helps to identify the most *feasible* intervention: an intervention that is not only suitable but also safe and effective, given a certain stakeholder field. We shortly introduce the steps to intervention and stakeholder analysis below and then further elaborate intermediate steps 1 and 2 as part of the Intervention Analysis in the remainder of this chapter.

Step 1: identifying thinkable interventions

Given the identified area(s) of intervention after the context analysis, there is a body of knowledge in the sector available that outlines a set of thinkable interventions in the identified area of intervention, referring to all potential alternatives for action related to the domain of intervention. This body of knowledge can be based on own field experience, the expertise of sector/cluster experts, evaluations of previous aid interventions, lessons learned papers and other documents outlining options for interventions or information about the conditions for success for such interventions. For example, if food shortage is considered to be a core problem to be addressed in a particular humanitarian crisis, there is ample documentation (such as Maxwell *et al.* 2013a, 2013b) showing that this can be done in various ways, ranging from free food distributions, cash-based interventions (such as vouchers), seeds and tools programs or livestock support.

Step 2: identifying suitable interventions

The second step is related to the question of how to decide which of these thinkable options is appropriate to use in a particular humanitarian context, i.e. how to arrive at the most suitable intervention(s) in this particular context. In other words, given the characteristics and causes of the humanitarian problem at hand (i.e. here: food shortage) and the context in which this problem is taking place, what would be the best fitting intervention to improve the situation (i.e. achieving food security)? We recommend conducting this analysis first without considering the various stakeholders involved, like one's own aid organization or other stakeholders present in the setting. Put differently, this first analysis is based only on the identified needs and vulnerabilities. We recommend this for two reasons.

First of all, humanitarian aid is based on the principle of humanity, and it is thus first and foremost assumed to be based on needs. Hence, organizations should be willing to conduct such a needs-based analysis first in order to identify options most suitable to address the needs. Second, a stakeholder analysis can only be instrumental if it is clear what the suitable options for action are. If, for example, the mandate of the organization is the first point of departure in such an analysis, this might result in a supply-driven approach in which ready-made solutions are forced upon a reality, resulting in a risk of negative side effects. Instead of humanitarian problems that require suitable solutions, this can lead to solutions looking for 'suitable' problems (March 1994, Heyse 2007). In the end, aid organizations should be prepared to also consider the option not to intervene, because others are already on the ground covering the needs, or because the expertise is lacking to implement the most suitable option effectively.

The type of analysis that helps to generate the most suitable intervention(s) can be done with help of a tool that belongs to the category 'theory-based *ex ante* evaluation' (Astbury and Leeuw 2010). *Ex ante* evaluation is a method to analyze the potential intended, unintended and adverse effects of projects, programs or policies *before* the intervention is actually done. This method is advantageous in that it may help aid organizations to identify potential pitfalls of their project prior to its implementation, thereby reducing costs and risks of side effects.

Step 3: identifying feasible interventions

Of course it is not realistic to assume that humanitarian aid can take place without considering the relevant stakeholders. After all, they can facilitate or obstruct the successful operation of humanitarian organizations (Collinson 2002, LeBillon 2000). Therefore, the third and final step in the analysis is to arrive at the most feasible intervention option, meaning that it is investigated how the stakeholder field – and thus the goals, interests and power of other actors in the domain of intervention – relate to the identified suitable interventions. In other words: given the set of identified suitable interventions and the characteristics of the stakeholder field, what option is most likely to be (most) acceptable and has least risk to be obstructed – and thus is feasible and safe. In order to arrive at this analysis, a *stakeholder analysis* is needed, which will be further elaborated in Chapter 12.

Taken together, the theory-based *ex ante* evaluation of potentially suitable interventions in combination with the stakeholder analysis, will help to generate an intervention option that is likely to help a specific stakeholder (here: a humanitarian organization) to realize her goal, i.e. the improved situation of people struck by a crisis. This is not to say that the method of theory-based *ex ante* evaluation is superior to a stakeholder analysis; they are both necessary in identifying a suitable *and* feasible project option. Only then are humanitarian projects most likely to result in intended effects.

We now first introduce the idea of theory-based *ex ante* evaluation, then explicate two concepts central to this approach – mechanisms and context – before we proceed to outline the practical steps humanitarian organizations can take to arrive at a set of suitable interventions based on this method.

The value of theory-based *ex ante* evaluation for humanitarian action

The tool we outline in this chapter is inspired by the literature on 'theory-based *ex ante* evaluation' (Weiss 1997, Coryn *et al.* 2011, Treasury Board of Canada 2012). *Ex ante* evaluation – as opposed to *ex post* evaluation – means that *prior* to a project's implementation, those who are responsible for proposing and planning projects or interventions, have made the effort to thoroughly reason through why and how the proposed activities will lead to an improved situation of the target groups and what the risks of unintended negative effects could be. It is thus a prospective evaluation method (see also US Government Accountability Office 1990) which resonates with the principle of *Do No Harm*, since it contributes to assessing the quality of the proposed plan of activities and to identifying potential harmful effects of the planned activities.

Theory-based *ex ante* evaluation serves three main purposes in this book: first, to help outline in advance what specific processes humanitarian project activities are assumed to trigger (i.e. what is the explicit program theory); second, to assess if and how that will generate the intended effects; third, to evaluate the risk of unintended negative effects. Such an exercise can be focused on short-term outcomes as well as longer-term effects.

The approach is labeled 'theory-based' because the purpose of such an *ex ante* evaluation is to unravel the often implicit assumptions (i.e. 'theories') of stakeholders that propose certain projects or program activities. Policy-makers and project planners are often not aware of their assumptions of how they expect proposed interventions to trigger certain processes (also referred to as mechanisms) so that positive change is achieved. These assumptions behind proposals for project activities can be seen as hypotheses or theories of how the world works in the view of practitioners, therefore the label 'theory-based'. If these assumptions remain implicit, one cannot detect potential inconsistencies or gaps in these assumptions. Since incomplete, inconsistent or incorrect assumptions can lead to unintended side effects of proposed humanitarian activities, it is of utmost importance to outline and evaluate them explicitly prior to an intervention's start.

The assumptions underlying proposed interventions are often referred to as a *program theory* (Astbury and Leeuw 2010, Leeuw 2003, Hoogerwerf 1990). Other terms often mentioned in relation to this theory-based evaluation approach are 'theory of change', 'intervention theory' or 'implementation theory' (Connell *et al.* 1995, Weiss 1997). The many different meanings and definitions of these terms (see for overviews Vogel 2012, Stein and Valters 2012, Blamey and Mackenzie 2007) converge on the idea that policy-makers and project planners need to make their assumptions explicit as to how a specific intervention is supposed to achieve its objectives.

It is not only important to make explicit the assumptions behind a certain intervention proposal, it is also and especially necessary to then think through the quality and validity of these assumptions: why would the assumed processes be effective? While doing so, it is important to be aware of the fact that the context in which interventions take place influences how interventions will work out. To illustrate: it is known that free food distributions are less suitable in crisis contexts where markets are still functioning (Maxwell *et al.* 2008), because the freely distributed food might be sold in the market, thereby leading to decreased prices and market distortion. Hence, free food distributions are more suitable in some crisis contexts than in others.

This idea that some interventions work better in some contexts than in others, because context influences the way the assumed program theories will work, is related to the Context-Mechanism-Outcomes (CMO) approach, which is part of the so-called Realist Evaluation tradition (Pawson and Tilly 2004, see Figure 11.1). The CMO approach helps to answer the question: what interventions (mechanisms) work well (outcomes) for whom, and in what circumstances (context)?

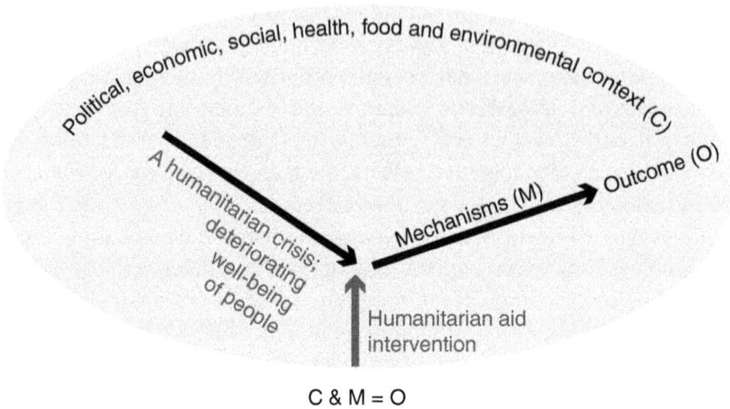

Figure 11.1 The Context-Mechanism-Outcome approach in humanitarian crises.

Translated to a humanitarian context, the question is: what specific aid activities (e.g. free food distributions or cash transfer) are likely to generate intended outcomes (e.g. increased food security) for beneficiaries, given the political, economic, social, health, food and environmental situation in a humanitarian crisis?

According to the CMO approach, an intervention is likely to fail or to generate unintended side effects if (1) the context conditions obstruct the assumed or intended program theory from working or even distort the program theory's working; and/or (2) the intervention itself does not trigger the assumed processes and mechanisms assumed by the program theory. In this way, the CMO approach can be helpful in outlining a method to identify prior to a project's start those context conditions that may hamper or facilitate particular humanitarian aid activities (based on existing knowledge and insights) and to find out why and how exactly certain program theories and activities might (not) work in particular contexts. Hence, this approach facilitates that more informed programming decisions can be made.

We will now proceed with elaborating the CMO approach by further clarifying how to think of interventions as triggering specific mechanisms that generate certain effects, and what the impact of context conditions can be. We will illustrate the general principles of this method with insights on emergency food interventions and then proceed with outlining specific steps as how to achieve a theory-based *ex ante* evaluation of a proposed humanitarian aid project.

Humanitarian aid as mechanisms-generating intervention processes

The literature on theory-based evaluation argues that any type of intervention aims at triggering certain mechanisms towards change or improvement of a situation or group. This also holds for humanitarian aid, which aims at positively intervening in the situation of human beings, as a way to improve their food, health and economic situation, and sometimes also their social and political situation. This can only be achieved if the intervention triggers certain mechanisms that lead to an improved situation. In this section we further elaborate what mechanisms are, why they are relevant to identify in humanitarian interventions and how they can be identified.

Mechanisms are reflected in the 'assumptions about how and why program activities and resources will bring about change for the better' (Astbury and Leeuw 2010: 364). Mechanisms are those elements that help explain *why* and *how* an intervention generates intended effects, as will be later illustrated in this chapter. Theory-based *ex ante* evaluation of proposed projects thus has an explicit explanatory dimension: it should provide arguments and evidence of how and why the program or project activities are expected to lead to a change in the target groups' situation or behavior.

Theory-based evaluation proponents argue that interventions in the social domain – such as policy programs, social care interventions or humanitarian aid projects – are often measured in terms of outcomes or effects only, thereby treating the process from intervention to outcome as a black box (Astbury and De Leeuw 2010). This prevents the contribution of the intervention to the outcome from being separated from other factors that may also contribute to certain outcomes (such as sheer coincidence).

There are currently methods in place that facilitate outlining the causal chain between interventions and activities and the outcomes they aim to generate (i.e. logframes). However, these methods often insufficiently explicate the assumed causal relationships between an intervention, the response that the intervention generates and the outcome that is then achieved (Gasper 2000). The logical framework, for example, asks for a systematic listing of project objectives, a set of indicators that would make these objectives measurable, and a set of assumptions that ideally would reflect the project theory and the conditions in which this project theory is likely to be successful. These assumptions are however often not described in detail but presented as general risks that could hamper the project, such as sudden fighting, so that food convoys are blocked. They often do not really reflect an intricate project theory of how the aid activity is going to contribute to an improved situation of beneficiaries and which context conditions are expected to hamper or facilitate the project's working. Consequently, the results of such methods are often descriptive rather than explanatory, and they provide little insight into the mechanisms triggered by the intervention (Coryn *et al.* 2011).

One explanation for this lack of explanatory depth is that the concept of mechanisms and program theory is difficult to grasp and work with for practitioners (Weiss 1997, Coryn *et al.* 2011). However, some work has been done to help practitioners with grasping the concept of mechanisms in particular sectors. An example of this is the work by Shapiro (2006) who outlined theories of change related especially to conflict interventions. She distinguishes between interventions aiming at generating individual level change, at changes in groups or relationships and at change at the macro level (i.e. structures, systems and institutions). Also humanitarian aid can be aimed at generating change at these levels, with the individual and group level being the dominant focus.

For each level of change, Shapiro outlines examples of mechanisms that can be triggered to generate change by means of interventions. For example, she elaborates individual-level change mechanisms in terms of cognitive, affective (emotional) and behavioral change. Cognitive change can be triggered when people learn new things, grow awareness, or are able to reframe information. Affective change can be triggered by emotional catharsis, whereas behavioral change can be triggered through improving certain skills. At the group level, change in relationships can be established by altering or enlarging people's networks, whereas change in macro structures can be related to judicial and legislative reform but also to altering economic and social structures.

The above insights can also be applied to humanitarian aid: part of humanitarian work is aimed at the direct physical improvement of people by giving them health care, water or food. Another part of aid is aimed at changing the structural circumstances people live in, for example, by providing them with money or resources to improve their living conditions, such as cash, cooking material and shelter. A third part of humanitarian aid activities is aimed at achieving individual or group level change, for example by means of training and information sharing, as is often done in water, sanitation and hygiene projects or livelihoods interventions.

Example of mechanisms triggered in emergency food interventions

We will now illustrate what we mean with the term 'mechanism' in a humanitarian crisis context by focusing on the domain of emergency food interventions. What would be, for example, the mechanisms generated by free food distributions and by cash transfers, both popular food aid interventions? The mechanism for free food distributions is quite straightforward: by giving people in-kind food for free, the assumption is that they can immediately consume this food, so they get a higher calorie intake, which directly addresses malnutrition. It is thus a biological/physical mechanism that is triggered. In the case of cash transfers, a different mechanism is triggered: first people get money so their resource base expands (their structural circumstances improve), and it is assumed that an expanded resource base will help people to address their malnutrition, for example by going to the market to buy food, seeds or cattle.

By making these assumed mechanisms explicit, one probably is immediately inclined to add critical remarks because these assumptions might not be correct, consistent or complete. For example, why would the simple act of giving people money lead them to buy food? This assumes first of all that lack of money is the problem of their malnutrition and not something else, such as unsafe conditions to go the market or lack of supply on the market. What is the evidence for this assumption? This is something that can be investigated. Second, the assumption seems to be that one is so hungry that one would spend the money on food and not on something else, such as paying off debts, medicine or school fees. The same goes for the assumed mechanisms triggered by free food distributions. Such an intervention assumes, for example, that one is so hungry that people will eat what is given to them and will not use it for other purposes, such as selling the food on the market. There are thus quite a number of assumptions related to the likelihood of success of cash transfers and free food distributions.

Each proposal for a humanitarian activity is thus grounded in a set of assumptions of how the activities will generate the intended effect. If one of those assumptions does not hold (for example, money is not the issue but cash transfers are given), the activity will most probably not generate the intended effects or even have negative side effects. It is thus crucial to make these assumptions explicit and to relate them to the context in which the activity is planned to take place, in order to analyze whether the proposed activities are likely to work as assumed. This exercise to unravel the mechanisms of proposed humanitarian aid activities can be started by asking the question: 'why and how will project/program activity Y generate intended outcome X?' The next step is then to critically assess whether this question is answered completely, consistently and/or correctly. We will later in this chapter specify the steps for doing this.

Humanitarian aid as context dependent processes

Being aware of the different mechanisms that certain interventions are assumed to trigger, is useful to think through in what situation to opt for what intervention. It is here where knowledge of the context of the humanitarian crisis – as suggested in Chapters 3 to 10 – is crucial. The context part of the thought experiment relates to the question for whom the intervention might work and in what specific circumstances (Pawson and Tilly 2004). In the case of humanitarian interventions, we propose to answer this question with the help of information about the social, economic, political, food, health and environmental context.

To return to our example, whether either free food distribution or cash transfer is a suitable intervention in case of malnutrition – and thus will trigger the mechanisms that are assumed to operate – strongly depends on the context in which one aims to intervene. Based on lessons learnt from previous humanitarian experience (see, for example, Harvey and Bailey 2011, Maxwell *et al.* 2008), we know that cash transfers as means to address malnutrition are most likely to be successful if they are implemented in situations where, amongst others, there is an operational banking system, and markets are accessible and functioning properly in terms of sufficient supply and choice. If cash is transferred and markets are inaccessible, for example due to violence, then it is likely that the cash cannot be used and people remain malnourished. If markets lack supply, the increased demand in markets due to the fact that people have more money to spend, might create a risk of rising prices and scarcity so that people might not be able to buy sufficient foods to improve their food situation. Hence, the economic context in terms of market characteristics is of utmost importance in deciding on a specific set of activities aimed at improving food security.

Free food distributions are arguably more likely to address malnutrition successfully in situations of inaccessible or badly operating markets, and in case of acute malnutrition (Maxwell *et al.* 2008). Nevertheless, other factors related to the context of the intervention need to be taken into account. First, the type of food should be carefully decided on because it otherwise might not be eaten or will be sold. This requires knowledge of local diets and staples. Second, the quantity of food should be carefully decided on, taking into account the norms and customs of food distribution practices in the household. For example, sometimes food is not equally distributed in the household, so that the family members lowest in the hierarchy might not receive the quantity of food individually needed to reduce their malnutrition (Debevec 2011). Hence, distributing food on the basis of a standard for minimum calorie intake per person might not result in improved nutrition of all household members. The above example shows that detailed knowledge of the economic context (functioning of markets), health context (type of malnutrition), food context (availability of and preferences for food), political context (safety issues) and social context (family norms) are crucial to determine what type of activity to employ to address malnutrition.

Such detailed knowledge of the context is all the more important since humanitarian crises can be typified as combining elements of both complicated and complex intervention contexts (Roger 2008). Complicated intervention contexts are characterized by multiple components and a plurality of stakeholders involved, creating coordination challenges. In addition, multiple mechanisms might be simultaneously necessary to trigger processes of change. For example, in case of cash transfer interventions, training might also be required. Moreover, across contexts different combinations of mechanisms might be essential to achieve success.

Complex intervention contexts are characterized by interconnected feedback and learning processes, with reinforcing or obstructing loops that can lead to disproportionate effects or the lack of any effects. The effect or outcome of an intervention is thus very much interdependent on these interactive feedback and learning processes at work, so that effects or outcomes emerge out of these complex dynamics. In these contexts, interventions will never work as straightforwardly as it may seem from the onset. Even if an intervention worked well in one context, this is not to say it can be expected to be equally successful in other contexts. Cartwright (2012) describes a striking example of how similar interventions impact differently in different contexts. She discusses World Bank nutritional interventions in Bangladesh, Ethiopia and Uganda based on a report by Save the Children (2003). The Bangladeshi project was modeled after a project in the Indian Tamil Nadu state, but whereas the Indian project was successful, the Bangladeshi project was not. The project consisted of a combination of providing supplemental feeding to children and increasing the mothers' knowledge about food. One of the factors that explained the different success rates had to do with different customs of how food was distributed in the household. In Bangladesh supplementary food for the child was often passed on to another household member, preventing the child from having sufficient calorie intake. Another explanation was in the assumption that increasing the mothers' knowledge would be beneficial for the child's nutritional status, but in the Bangladeshi context mothers do not have the authority to decide about their children, but the men and the mothers in law do (Cartwright 2012).

Hence, especially in humanitarian crises a careful and conscious identification of mechanisms and context conditions (i.e. CMO configurations) is required. Of course such an *ex ante* evaluation of proposed interventions will not guarantee success in interventions. But it will help to at least identify incomplete or incorrect program theory, and to signal risks for unintended side effects of interventions.

Context-Mechanism-Outcomes (CMO) thinking for humanitarian aid

The outcome of a humanitarian project (i.e. improvement of the target group's situation) is the sum of the mechanisms triggered by the intervention, given a specific context. The next sections specify the steps to arrive at suitable intervention options for the humanitarian sector.

Step 1: from evidence to priorities: identifying areas for intervention

The product of the Comprehensive Context Analysis is a radar graph summarizing the most important information on vulnerabilities and capacities for a particular humanitarian crisis. The radar graph allows the identification of domains (e.g. food, health) in which interventions are most needed (usually domains where vulnerabilities and needs are high and capabilities are low). The domains of intervention can be identified in terms of the type of context (social, political, food/health/water, environment, economics) but also in terms of UN clusters: food, nutrition, shelter, WASH, education, etc. Based on the Comprehensive Context Analysis one or more areas for intervention can be prioritized, which in our particular case refers to food.

Step 2: from priorities to programming ideas: generating thinkable intervention options

After having prioritized one or more areas of intervention, the aim is to work towards an intervention plan, for example in terms of a humanitarian aid project proposal. How do we arrive at such a plan in light of the proposed method of *ex ante* theory-based evaluation? A first step is to generate a list of thinkable intervention options. There is ample experience and expertise in the humanitarian sector that humanitarian aid staff can use to derive such a list from. An important element in humanitarian programming is thus to ensure that one generates a set of thinkable alternatives for action for a certain intervention domain. These alternatives can be generated by means of sector and cluster experts one can turn to, by using own field experience and by studying documents that outline lessons learnt from previous experience such as found in evaluation reports, good practice reviews, best practice papers or case studies of success and failure. There are many sources one can turn to, such as ALNAP, HPN/ODI, information provided by the UN clusters and UN agencies, and academic insights.

With regards to our example of food interventions, there are quite a number of publications, documents and insights one can draw up. Box 11.1 outlines a few examples.

Box 11.1 Information sources about emergency food interventions

Examples

Action Contre la Faim (ACF) International, *Introduction to Food Security Intervention Principles.*

ALNAP (2011) *Humanitarian Action in Drought Related Emergencies*, Lessons Learned Paper.

Levine, S. and Chastre, C. (2004) *Missing the Point: An Analysis of Food Security Interventions in the Great Lakes*, Humanitarian Practice Network Paper No. 47.

Maxwell, D., Sadler, K., Sim, A., Mutonyi, M., Egan, R. and Webster, M. (2008) *Emergency Food Security Interventions*, Humanitarian Practice Network Good Practice Review No. 10.

Maxwell, D. Stobaugh, H., Parker, J. and McGlinchy, M. (2013) *Response Analysis and Response Choice in Food Security Crises: A Roadmap*, HPN Network Paper No. 73, London: Overseas Development Institute.

Save the Children (2003) *Thin in the Ground: Questioning the Evidence behind the World Bank-funded Community Nutrition Projects in Bangladesh, Ethiopia and Uganda*, London: Save the Children UK.

Based on these multiple information sources one can extract a set of thinkable interventions, such as outlined in the middle column of Figure 11.2. The same can be done for other intervention domains, such as shelter, nutrition or WASH.

Depending on time constraints, the duration of this step can be flexible, but the main point is that a number of alternatives for action are generated based on existing experience and expertise in the field. By purposefully pursuing the generation of alternative thinkable actions, this increases the likelihood that alternatives are proposed that might not have been thought of at first instance. It thus facilitates that one looks beyond the known or often-used options and thereby helps to overcome tunnel vision.

Step 3: generating suitable intervention options: CMO thinking

Based on the set of generated thinkable interventions, a limited number of suitable interventions can be selected, i.e. interventions that match the specific problem and context at hand. For each of them, it is necessary to specify the

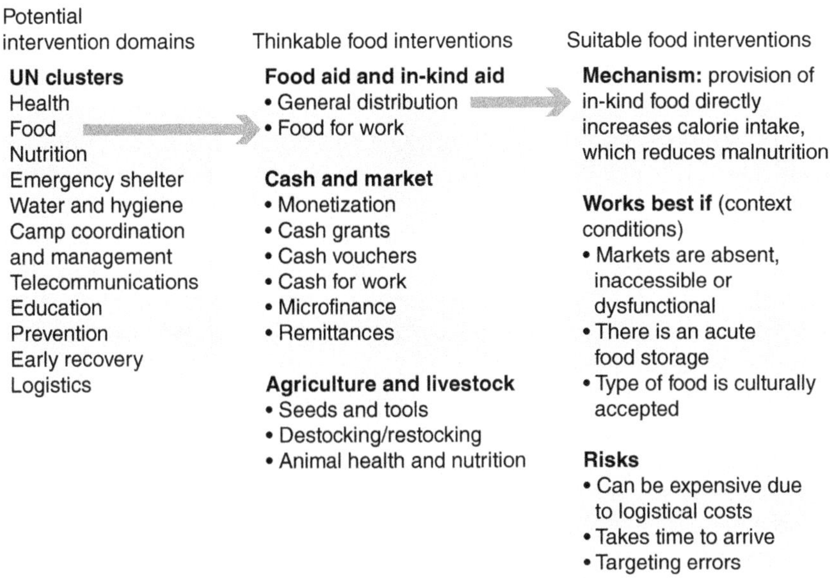

Figure 11.2 From prioritized intervention domains to thinkable and suitable food security interventions (sources: based on Maxwell *et al.* 2008, 2013a, 2013b and ACF).

mechanisms and the context conditions under which these interventions are likely to fail or to succeed.

Identifying mechanisms: constructing program theories

Mechanisms can be explicated by constructing a program theory, for example by formulating a set of IF–THEN statements. Questions that could be asked are:

- What mechanism(s) is/are a specific activity in an intervention assumed to trigger? One can use a structure of if–then statements to answer the question. For example, IF activity X (free food aid) is employed, THEN this will lead to outcome/effect Y (increased malnutrition) BECAUSE the calorie intake of individuals will go up immediately.
- Does each activity trigger one or more mechanisms? If so, do these simultaneous mechanisms contribute to the intended outcome or might they obstruct each other?
- To what extent do activities interact with each other? Do they 'flank' (i.e. strengthen) each other or do they obstruct one another in relation to the intended outcomes/effects?

Let us take emergency seed aid interventions as an example of a potentially suitable intervention method in a given humanitarian crisis. What would a program theory for an emergency seed intervention look like? Using IF–THEN statements, one could come to the following reasoning: IF one distributes seeds to targeted households, this will THEN give households the means to plant (more) crops (M1 in Figure 11.3), which is THEN expected to lead to more crops (M2) that can be harvested. IF the crops can be successfully harvested (M3), the harvest can THEN either be consumed immediately (M4a) or sold on a market (M4b), or a mix of both. IF one eats it directly, THEN the calorie intake of households will directly improve, so that the nutritional status of the household members will improve. IF the harvest is sold on the market, THEN this gives the household the opportunity to earn (more) income with which they can buy foods on the market (M5). Figure 11.3 summarizes the above IF–THEN statements in terms of mechanisms (M1–5).

Questions that come to mind on the basis of the above are:

- What is the underlying cause of nutritional problems? Is this related to lack of seeds or to other issues such as food availability, food access or, for example, health-related problems? In other words: to what extent will this intervention trigger mechanisms that address the causes of the nutritional problems?
- Why and how should increased availability of seeds result in increased yields?
- How does increased availability of food (harvest) lead to sufficiently increased calorie intake of household members to improve their nutritional status?

Ideas about which mechanisms to look at can be generated on the basis of documents, prior research, asking experts and stakeholders, and by logical reasoning (Weiss 1997). Criteria that can be used to decide to include or exclude a

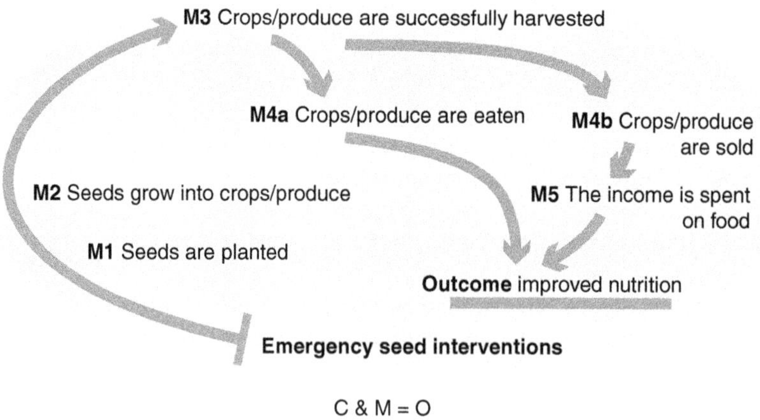

Figure 11.3 Crucial mechanisms behind successful emergency seed aid interventions.

mechanism in the analysis are: the assumptions of the program stakeholders, the plausibility of the mechanisms (how likely is it that these mechanisms will occur) and the centrality of the mechanisms to the program (Weiss 2002). Furthermore, it is important to critically reflect on the completeness of the assumed mechanisms, the consistency and coherency between mechanisms and the correctness of them.

Identifying context conditions

Once the assumed mechanisms have been outlined and reflected on, the next step is to connect these assumed mechanisms to the context characteristics to see how they relate to each other. For this step in the analysis, one can utilize the typology of contexts used in Part I. In the case of seeds interventions, the following context conditions are of importance for these interventions to be successful (based on Dijkhorst 2011, Sperling and McGuire 2010, Remington *et al.* 2002). We illustrate these context conditions in Figure 11.4.

- Food and health context. Seed interventions can only be implemented in not so acute malnutrition situations, since it takes time for seeds to grow and be harvested. For this intervention to work, target groups should still have other ways to uphold their food security situation, otherwise the increased food availability will come too late or the seeds will not be planted but eaten. Timing is thus the issue for a seeds intervention to be successful (C3 in Figure 11.4).
- Environmental context. These interventions will only work if growing crops at home is possible, i.e. there is space and there are favorable climate conditions (enough rain, fertile soil, adapted to the seasonal cycles, etc.). It is also crucial

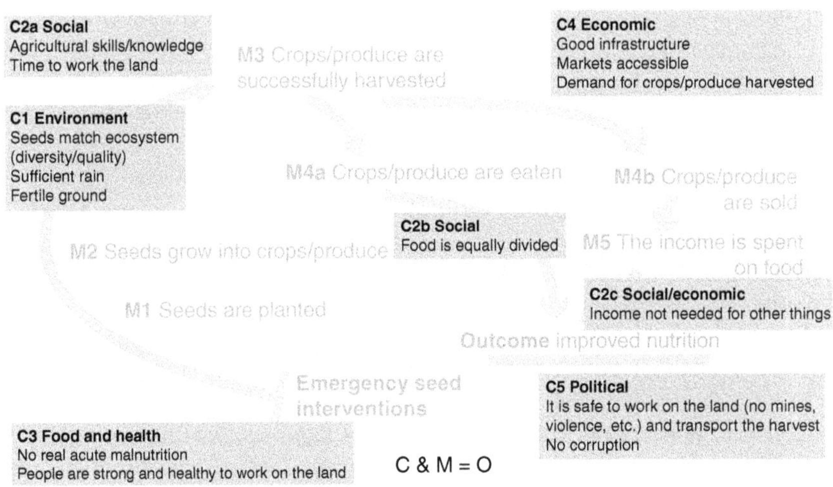

Figure 11.4 Required context conditions for successful emergency seed aid interventions.

that the types of seeds distributed match the environmental context in order to prevent the spread of diseases. Hence, distributing seeds is not enough; if they are of the wrong kind, the climate is not favorable, etc. there is a risk that the harvest will fail (C1 in Figure 11.4). A failing harvest will mean a failed increase in available food and the assumed program theory will not work.

- Social context. Household members need experience and expertise to grow crops themselves, and they should have time to do so (C2a). If knowledge, experience and skill are lacking there is a risk that mistakes are made and the planted seeds will not develop into a rich harvest. A failed harvest will not allow households to address their malnutrition. After the food is harvested and consumed, malnutrition for all in the family will only decrease if food is shared equally in the household (C2b). If one would plan to sell the harvest, the income should not be needed for other purposes such as paying off debts, because otherwise the money will not be spent on food (C2c)
- Health context (C3). Those who have time and skill to work on the land should also be strong and healthy enough to be able to work on the land.
- Economic context (C4). The distributed seeds should match the demand for food on markets. For example, distributing one type of seeds to a large group could promote a mono-culture which might in certain circumstances create increased competition on markets and increased vulnerability to shocks at the household level (due to crop illness, drought, etc.) due to a lack of diversification of produce. In case the intervention aims at stimulating farmers to sell their produce, there should be functioning markets nearby that have a demand for the produce harvested as well as a good infrastructure to transport the products. If these conditions are not fulfilled, farmers will not be able to either sell their produce on the market or get a good price for it, which will make the program theory less effective.
- Political context (C5). These interventions can only work if it is safe to work on the land, i.e. no risks of attacks, violence or landmines. Corruption could also distort the program theory to work as expected.

Towards Context-Mechanism-Outcome (CMO) configurations

Once the mechanisms have been explicated and the context conditions for successful intervention have been defined, and these have been compiled in graphs such as Figure 11.4, we have created a so-called CMO configuration. This configuration summarizes the key mechanisms underlying a successful intervention, based on what we know from existing publications and expertise, and the necessary context requirements for the intervention to work.

The CMO configuration can be compared to the Comprehensive Context Analysis of the particular humanitarian crisis at hand, so that one can establish whether the assumed mechanisms are likely to be triggered in this particular context. The key question is then whether the context conditions required for a successful intervention are present in the particular crisis.

In our example, it could be that the identified health and environment context conditions are present, i.e. people are healthy enough to work, the seeds match the ecosystem, the land is fertile enough for the seeds to grow and there is sufficient rain. In Figure 11.5 this is represented by a tick sign in C3 and C1. However, in this fictitious case, the social conditions are not met, because the context analysis showed that people miss skills and knowledge to work the land effectively (C2a), food is not equally shared in the household (C2b) and many people have debts (C2c), so they are likely to spend their new income on paying off debts instead on buying food. This analysis points out that there are serious risks to the seeds program, which is symbolized with the crosses in Figure 11.5. The program aims at decreased malnutrition, but this might be hampered by the above-mentioned social factors. One could thus already identify in advance groups and individuals for whom the intervention might not work that well: those who lack the knowledge to effectively work on the land, those household members who will not receive sufficient food due to inequality in food sharing practices and those households that need to spend the newly generated income on other matters than food. In addition, not all political and economic factors are met, so this asks for an additional analysis of which elements in these contexts are obstructing the program's aims.

Based on this analysis as visualized in Figure 11.5, one can start to consider whether some of the context conditions can be influenced by means of additional aid activities. For example, if people lack agricultural skills and knowledge, the intervention might need to be extended with a training element. However, not all context conditions can be easily influenced. For example, it will be difficult to change a hierarchy through which food is distributed in the household. Complementing the seed aid intervention with supplementary feeding for the weakest in

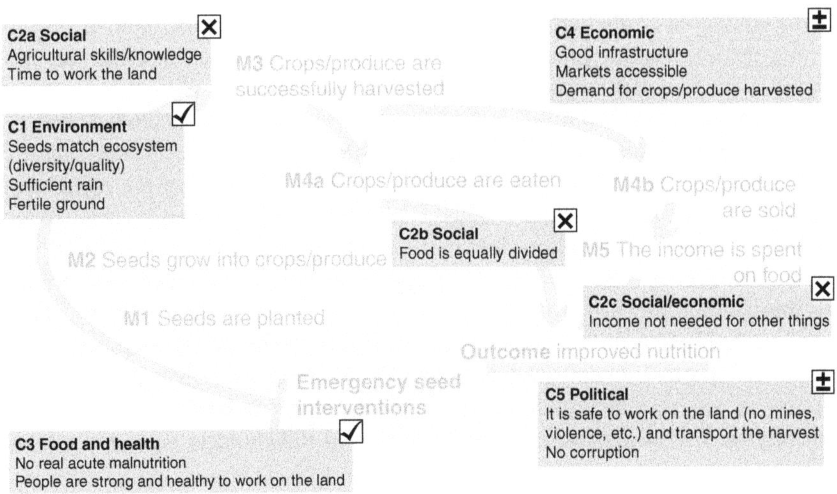

Figure 11.5 Fictitious analysis of a CMO configuration applied to a particular crisis context.

the household may be one way to address this problem. Moreover, sometimes the conclusion can even be that the particular intervention is not suitable in the current context at hand, for example if the land is not safe to work on because of landmines.

Now all steps have been taken in the Intervention Analysis, based on the method of *ex ante* theory-based evaluation, a quite intricate picture is available about the potential success of the planned intervention. The main activity in the project is the seeds distribution, and so far one could conclude that this distribution might make sense for some groups in this fictitious case. However, for others, one can already conclude that the distribution entails risks that can be partly addressed by adding to the main project activity a training program for those who lack agricultural knowledge and skills, and maybe a supplementary feeding program for those who receive least food due to unequal food sharing practices. Nevertheless, more insight in the political and economic context is needed to make sure that there are no barriers in these contexts that prevent people from, for example, successfully selling their produce on the markets or from working on the land due to safety problems.

Conclusion

This chapter aimed at outlining a method to evaluate in advance whether proposed humanitarian interventions can be assumed to generate the intended effects. The framework of theory-based *ex ante* evaluation was used to outline how to do this. Two steps were proposed: (1) to generate a set of thinkable interventions, and (2) to generate suitable intervention options by means of (a) reconstructing a proposed intervention's program theory and mechanisms and (b) to connect these assumed mechanisms to context conditions for success. The result of this thorough experiment is a Context-Mechanism-Outcome configuration that can be compared to the outcomes of the Comprehensive Context Analysis. Connecting the intervention plans to the context analysis is crucial for three reasons. First, it helps in determining whether the required context conditions are present and the proposed intervention is thus likely to generate the intended effects. Second, it helps to discover whether adjustments or complementary activities are needed because certain context conditions are not met (if these are conditions that can be influenced by aid activities). Third, it might even lead to the decision that the proposed intervention does not suit the particular context, or that one cannot influence crucial barriers in the context, and thus one would not initiate this intervention.

However, before the definitive decision can be taken that a proposed intervention should be implemented, one not only needs to know whether the intervention is *suitable* but also if the intervention is *feasible*, i.e. whether the intervention will be successful given the involved stakeholders. This is the topic of the next chapter.

12 Stakeholder analysis

Towards feasible interventions

Rafael Wittek

Introduction

In the previous chapter, we discussed a method to arrive at suitable intervention options. Before one can initiate a project, one more step is necessary: assessing which suitable interventions are also feasible given the stakeholder field. This chapter presents a tool for choosing feasible interventions from a set of suitable options.

Humanitarian interventions are usually carried out by humanitarian organizations. In order to successfully carry out their mission, these organizations depend on many other stakeholders. This can be other aid organizations, local governments, companies, political parties or any other influential player at home or in the area of intervention. Some of these players may turn out to be crucial allies in getting a project implemented, whereas others may effectively impede it. As a result, a mission's success crucially depends on the organization's ability to deal with relevant other stakeholders. This chapter introduces a tool that helps humanitarian organizations to do this.

Stakeholder analyses are frequently used in the humanitarian sector (e.g. DFID 1993, 1995, ODI 1999, Schmeer 1999, 2000). They are often required for funding applications, and there are many different ways to conduct them, with some frameworks focusing mainly on the identification of stakeholders and their interests and others proposing more complex analytical steps related to the interdependence between stakeholders. Stakeholder analysis is also a field of scholarly investigation, particularly in the areas of collective decision-making and the study of negotiation processes (e.g. Brugha and Varvasosvzky 2000, Bryson 2004, Mitchell *et al.* 1997, Savage *et al.* 1991, Stokman *et al.* 2013).

During missions, a stakeholder analysis becomes particularly useful in situations of competing stakes or disagreement between two or more actors, or when a collective decision has to be arrived at. A stakeholder analysis is usually carried out on behalf of a specific stakeholder, like a humanitarian organization. It can serve a variety of different purposes. First and foremost, it can help a stakeholder in achieving her policy goals. Second, it can help a particular stakeholder to systematically disentangle the different issues at stake regarding suitable interventions identified previously. Third, it helps to identify the players who have a stake in the issues involved, and to get insight into the relative effectiveness of different strategies to realize preferred decision outcomes. And last but not least, it is also well suited in situations where decision-making involves organizational security issues.

Stakeholder analyses help to answer questions like: What are the key issues and bottlenecks of an intervention? Who are the key stakeholders involved? How powerful are these stakeholders? What are their preferred outcomes on an issue? How important is a specific issue to particular stakeholders? Which stakeholders are potential 'allies' who would be willing to support your own organization in realizing your suitable intervention? These insights can then be used to craft influence and bargaining strategies that help the stakeholder to be more effective in realizing his or her preferred policy outcomes – i.e. to move from a suitable to a feasible intervention.

A stakeholder analysis can be seen as an analytical tool for the reduction of social complexity. Though it results in a highly stylized representation of social reality, this tool and its visualization opportunities have nevertheless proven to be an extremely powerful aide for those who want to get a structured overview over the stakeholder field. For didactic purposes, it is useful to disentangle the tool into eight cumulative steps: issue definition, stakeholder identification, stakeholder description, definition of outcome continuum, stability analysis, stakeholder classification, description of negotiation landscape, relationship analysis. These steps will further structure this chapter, but before presenting them, we discuss the importance of expert information in stakeholder analysis. The section thereafter will briefly introduce a hypothetical example case. In the eight subsequent sections, this case will be used to illustrate each of the eight steps in a stakeholder analysis. The chapter concludes with a discussion of the scope conditions and limitations of the tool.

The importance of expert information

Carrying out a good stakeholder analysis requires the collection of reliable information on the preferences and constraints of all stakeholders. Often, this type of information cannot be gathered from each stakeholder directly, since the party conducting the stakeholder analysis may be in competition with one or more of the involved stakeholders. Therefore, stakeholder analyses usually rely on 'experts' to provide estimates of how specific stakeholders view the different issues. Who is an expert depends on the specific situation. It is a knowledgeable individual who has sufficient background information about one or more stakeholders, their context and the problematic issues at stake. Often, such experts are available in one's own organization. But also local actors (journalists, members of NGOs, researchers) usually have considerable insight into the situation. Much useful information on some stakeholders' positions may also be gained from scanning the public media on relevant statements.

Some of this information, but also information provided by experts, might be contradictory, and resolving these contradictions is a major precondition for all subsequent steps in the analysis (see also Chapter 4). Contradictions may have many reasons. For example, informants may have a stake in some of the issues themselves, and may therefore either deliberately or unintendedly misrepresent some information. Likewise, not all informants will be equally well informed about all stakeholders' position, salience and power on all relevant issues. The person or team carrying out the analysis should invest some effort into resolving these inconsistencies before proceeding with the analysis. Since quantification of estimates is an important element of a stakeholder analysis, some effort should be put into how to systematically elicit information from expert(s) during an interview session (see below).

Hypothetical case

After having conducted a comprehensive context analysis, a large Christian Humanitarian Healthcare NGO (CHH) has identified an emergency situation in a remote rural area in Africa. The NGO is already present in this country, but not in this particular area. Members of the NGO's country management team have identified health to be a severe problem in this remote area, based on several indicators as outlined in the chapter on the health context. The health problems are partly due to the fact that the communities in this area are hosting refugees from a neighboring region that is suffering from violence of a rebel group. Many have fled into this remote area and relocated to small villages, where the receiving communities welcomed them warmly. However, this influx of refugees has led to an increasing demand for food, which is not readily available. At the same time ongoing drought threatens agricultural production. Malnutrition is on the rise and is likely to increase further. Due to the fact that now many people live closely together, water and sanitation is also becoming a problem and related diseases are spreading.

By conducting an *ex ante* evaluation of potential suitable ways to address the health problems in the area, the organization has identified the following set of suitable health interventions. First, it seems necessary to support the few local health centers and clinics in order to deal with the increase in numbers of patients due to the refugee influx. These centers and clinics need extension of services and staff in order to help those who flew from violence and got injured (i.e. offer surgery facilities), next to those who suffer from malnutrition (i.e. offer therapeutic feeding facilities) and diseases related to decreasing hygiene and sanitation (i.e. diarrhea). Second, a water and sanitation instruction program seems necessary to inform the local population about the risks of living so closely together in terms of water and hygiene, and how to avoid associated diseases. Third, next to health centers and clinics, mobile health teams should be composed that can travel to patients who cannot travel to a clinic in the most remote areas in the region.

CHH has also detected through its context analysis that the host government has so far not allowed any international aid organization to enter the area and to provide aid. One local NGO is operational in the area, and is of the same denomination as the government (which CHH is not) but is lacking the resources and expertise to tackle the situation. The government seems to see it as a sign of their own weakness if it would allow international actors to enter the area. However, the government clearly does not have the resources to extend the health services in the region. CHH believes that all three activities – as outlined above – are necessary in order to effectively address the deteriorating health situation in this area. In order to find out whether there are any openings to initiate these three activities simultaneously, the country management team carries out a stakeholder analysis.

A stakeholder analysis consists of eight major steps, most of which can be carried out consecutively (the following description of each step draws on and expands the short but highly useful practitioner oriented article by Allas and Georgiades 2001). (For a more in-depth discussion of the scientific background behind the whole approach, see the work of Stokman *et al.* 2013.) The first two steps – issue definition and stakeholder identification – are of course strongly intertwined, since issues arise due to (conflicting) interests between different stakeholders, and the definition of issue requires some background knowledge concerning which stakeholders are involved. At the same time, when identifying stakeholders, it is important to keep in mind which issues are relevant: though there may be a large number of potential stakeholders involved in a complex emergency, only a small subset of them might be relevant for a specific decision situation. Hence, the first two steps may best be conducted simultaneously.

Step 1: issue definition

Main elements

Any stakeholder analysis stands and falls with the quality of the first step: the identification and definition of relevant issues. An issue is relevant for a stakeholder analysis if it is controversial. That is, at least two parties diverge with regard to the decision outcomes they prefer concerning this issue. In relation to intervention design, which is the main focus in this chapter, an issue represents a set of suitable intervention options as identified by an aid organization, of which one knows that controversy exists with other parties, such as local authorities, rebel groups, village communities or fellow agencies.

Issues can also be related to controversies internal to an aid organization. A simple example for such an issue could be a conflict within the project management team of a mission about the percentage of the budget for a mission that should be allocated to a specific activity. This could be reflected in, for example, the number of staff that should be sent out to a mission. Some members of the decision-making body may prefer not to be involved in the mission at all and opt for sending out nobody (preferred decision outcome: ‘zero’). Other members may favor full involvement of their organization, and opt for sending out the maximal number of staff (say 20). Finally, a third faction may be in favor of involvement, but prefer a more cautious allocation (say, ten staff).

Issues are defined correctly if they meet at least two conditions. First, all possible decision outcomes regarding it can be specified as concrete decision alternatives. In the example above, the range of possible decision outcomes is defined by the two possible extremes: a maximum number of staff that can be sent out, and no staff. In theory, any number (0–20) along the continuum between these two extremes represents a possible outcome.

Second, all decision alternatives can be ordered on a one-dimensional scale. In the example, the number of staff defines this dimension, but any other meaningful dimension is possible. For example, the allocation decision in the above example could be discussed in terms of percentages of the total budget, or in terms of maximal amounts of euro of a total budget. Other issues might involve outcomes like the amount of time (e.g. days) to be spent in a location. But it is also possible to scale issues containing 'non-numerical' outcome alternatives. For example, an issue relating to the degree of collaboration between two NGOs may contain the following qualitative option on the outcome continuum: no collaboration; partial sharing of transport facilities; full sharing of transport facilities. In such cases, one would still allocate numerical values to these options (e.g. 1, 50, 100, respectively).

It is important to keep in mind that the numerical representations are supposed to reflect the relative strengths between the different outcome alternatives. Hence, what counts is not only the fact that an outcome alternative is higher or lower on the scale, but also how close an alternative is to another one (see the issue specification below for an example). Note further that to facilitate analyses, it is useful to normalize the outcome continuum of all issues to a scale ranging from one to 100, with '100' representing the most extreme position taken by a stakeholder, and 'one' the lowest position taken by a stakeholder. This normalization step is presented in detail in step 3.

Often, what seems to be one single issue (e.g. set up a water and sanitation installation in a remote area) with a binary outcome space (yes vs no) in fact hides a multidimensional set of more complex issues. Since their conflict potential may vary, carefully disentangling such sub-issues often is a fundamental first step towards successful negotiations. In the water and sanitation example, sub-issues might be the location of the watsan (water and sanitation) installations and the type of watsan installations. For example, a local population might be used to a certain type of latrine, but these are not very resilient to storm or earthquakes. An aid organization may therefore prefer a more resilient installation, while at the same time it has to be sure that these facilities will also be used.

For a stakeholder analysis to be effective, it is also essential that all relevant issues are covered. Which issue is relevant and which one is not usually emerges during the issue definition phase. Usually, in order to unravel all relevant issues and the related outcome alternatives, it is necessary to also first identify the involved stakeholders. Hence, the first two steps (issue definition and stakeholder identification) in a stakeholder analysis are usually closely intertwined.

Case illustration

The country management team in our fictitious case decides to carry out a stakeholder analysis. To do so, the team first carries out some informal discussions with people in the NGO's professional network: direct colleagues working in a project in a neighboring region, government representatives and representatives of other NGOs in the capital. Based on these discussions, the team realizes that the desire to initiate the three health related activities simultaneously in fact touches upon at least the following four core issues that need to be taken into consideration if CHH would decide to enter into any lobbying and negotiation efforts with relevant stakeholders:

Size of the remote unit ('size')

Currently, the plan is to supply the remote area with 70 humanitarian workers, including physicians, para-medics, and support staff, of which ten locally hired. Consequently, two possible outcomes on this issue are zero staff in the case of failure to get access and '70' in the case full access is achieved. But it may also be possible that one or more of the involved stakeholders actually may prefer that even more staff is allowed to work in this region, for example to double its size.

Type of services ('services')

The health NGO ideally would like to initiate a range of medical services, as outlined above. For the moment, the country management team considers the support to health centers and clinics (on site services) as most relevant, compared to the mobile health teams that visit remote households. The outcome continuum could be specified as follows: (1) 'no services', (50) 'on-site only', (100) 'on-site and household visits'. Note that in this case, on-site visits and household visits are considered as part of one dimension of increasing 'intensity of services', because the NGO would not be able to carry out household visits without having a site in the remote area. Hence, household visits require the existence of a site, and would constitute an additional service. Note further that the distance between each pair of outcomes is chosen to be the same (50 points). This implies that the NGO considers the extra investment or effort that is necessary to add household visits to on-site services as equally strong as setting up and maintaining a new on-site facility. Of course it would be possible to adjust these figures if the estimates regarding the relative investments change. For example, if adding household visits would be perceived as requiring comparatively less investments once a site has been established, the analyst could opt for allocating the value '75' to the outcome on-site only. As a result, the distance between this outcome and the outcome on-site and household visits (100) would decrease to 25 points, whereas the distance from no services (1) to on-site only (75) would increase to 75 points, reflecting the fact that the relative investment for setting up a site is higher than the additional effort needed to add household visits to the set of services.

Degree of collaboration with other NGOs ('collaboration')

Since CHH is not used to working with local partners, it does not have any formal collaborative arrangements with the one local NGO in the area. Yet the government might prefer more cooperation with the other NGO operational in the area in order to reduce the visibility of the fact that the government cannot cope and needs international assistance. Also, the government might see cooperation as a way of controlling the international NGO, which amongst others, is of a different denomination than the government and the local NGO. Cooperation can of course take many forms. In this particular area, one option to cooperate with the local NGO would be to jointly implement the watsan hygiene instruction campaign. Given the budget constraints and other tasks with higher priority of both missions, setting up an effective campaign is only possible if part of a joint effort. The normalized outcome continuum here ranges from '1' in case no investments are made ('0$') and the campaign is not realized, to 100 (i.e. the sum of the budgets that are maximally available in both missions, let's assume 100,000$).

Duration of the health program ('timing')

It could be that the length of the project activities might matter for some stakeholders. Possibly the government would agree with a short-term emergency intervention (one month) or a medium-term intervention (six to nine months), whereas the Christian NGO and maybe also other stakeholders believe an 18 month intervention is needed to fully address the problems. Hence, the non-normalized outcome continuum for this issue spans any period between one and 18 months.

Step 2: stakeholder identification

Main elements

Along with the definition of the issues, the relevant stakeholders need to be identified. These are players who might influence decisions or their outcomes on the issues. Examples for stakeholders in humanitarian crises are (local) governments, local NGOs, other humanitarian organizations, representative bodies of the local population, but also other groups with power or influence (e.g. rebels). Note that since organizations often consist of several subunits with decision-making power and resources, a stakeholder analysis may require disentangling these units as different stakeholders (e.g. departments). Deciding whether to disentangle the units as own stakeholders is a matter of feasibility (the higher the number of stakeholders, the more complex the stakeholder analysis). The easiest way to reduce the number of stakeholders to be considered in a stakeholder analysis is to look at their goals and their decision-making procedures. If all goals and decision-making procedures converge, then the respective stakeholders can be 'merged' into a single one (e.g. all ministries could in some cases be represented as one stakeholder – the government). If all goals converge but decision-making procedures do not (e.g. the UN does encompass the UNDP, but the UNDP has its own independent governing board of state representatives), then we advise against merging them into a single stakeholder.

Case illustration

Based on the informal explorations, the country management team identifies the following five key stakeholders:

A The mission of the health NGO CHH ('Mission'), represented by the country management team, located in the capital of the country.
B Churches in the area, which are of the same denomination as the NGO and represented by a coordinating council representing them ('Churches').
C Government of the host country ('Government'). Located in the capital, close to the offices of the mission/country management team.
D The local NGO operational in the area ('Competitor'). This is also an NGO working in the medical sector, and providing limited but similar services. This local NGO is of another denomination than the international NGO.
E An association of village heads ('Villages'). This association represents the interests of all villages in the area of operation.

Step 3: stakeholder description

Main elements

After the stakeholders have been identified, four crucial characteristics of each stakeholder need to be determined: position, salience, power and effective power. We refer to this operation as the stakeholder description step. It consists of allocating numerical estimates for each stakeholder's position, salience, power and effective power for each issue. The estimates are derived from discussions with the experts.

First, a *position* denotes the stakeholder's preferred outcome on a specific issue. This position needs to be quantified using the outcome space that defines the possible decision alternatives for this issue. For example, stakeholder A's position on the issue 'budget for mission X' may be 50,000 euro, whereas stakeholder B's position may be 100,000 euro and stakeholder C's position may be '0'.

Second, it is necessary to estimate the *salience* of a stakeholder's position with regard to a specific outcome on a specific issue. The key question here is: how important is the issue to the stakeholder, compared to all other issues and compared to all other stakeholders? For example, would the stakeholder be willing to lose all other issues in order to realize his or her preferred outcome on this issue? A stakeholder's estimated salience is expressed on a scale from zero to 100. A salience of '100' means that the issue represents the most important issue for the stakeholder. A salience of '50' means that it is one of several issues that the stakeholder considers to be important, but is not the most important one. A salience of '0' implies that the stakeholder has no interest in this issue. The salience values reflect the relative importance of an issue – compared to all other issues in the analysis – to the stakeholder. This implies that in stakeholder analyses involving more than one issue, a single stakeholder will attribute maximum salience to only one issue.

Third, the *power* of a stakeholder with regard to a specific issue needs to be assessed by asking: compared to other players, how much power the stakeholder has to affect the decision on this issue? Again, an estimate can range from '100' for the stakeholder with the greatest power on an issue to '0' for the stakeholder with least power. Since this is a relative rating, there can only be one stakeholder who is allocated maximum power on a specific issue. The degree of a stakeholder's power can be based on formal decision-making power, material power (money, facilities, other resources) or immaterial influence (e.g. expertise or ability to mobilize public opinion). For example, in organizations, a workers' council may have the formal right to give advice in some decisions, but might need to give its consent in other decisions, like reorganizations. At the same time, workers' councils may have much informal power to enforce some outcomes, e.g. through influencing the opinions of employees. An extreme form of power is coercion based on (physical) violence.

Fourth, based on a stakeholder's salience and power, it is now possible to calculate a stakeholder's *effective power* as a product of both salience and power (divided by 100 for the resulting variable to stay in the range of 1–100). The reasoning behind this multiplication is that if stakeholder A has no interest in an issue, he or she is also indifferent about the possible outcome on this issue. Consequently, such a stakeholder is unlikely to invest much effort into realizing this outcome.

Case illustration

With the help of the experts in her network, the country management team carries out the stakeholder description step. For each of the five stakeholders, she determines, per issue, their position, salience, power and effective power. For the purposes of this example, we limit the description to the first issue, the size of the unit (see Tables 12.1 and 12.2).

Step 4: outcome continuum

Main elements

With the first three steps carried out, it is now possible to summarize the data in matrix form and enter it into a spreadsheet (see Table 12.2). For each issue, the rows contain all stakeholders, and the columns contain each stakeholder's position, salience and effective power (salience times power) for this issue. This data enables calculation of a theoretical compromise position for each issue, i.e. a hypothetical outcome that could come about by a simple vote without negotiation. There are of

Table 12.1 Stakeholder description for issue 1: 'size of remote unit'

	Position	*Salience*	*Power*
Mission	Prefers to go ahead as planned with 70 staff or even more. Considers a 10% decrease in size as the critical threshold to ensure quality of the services and guarantee safety (70).	Compared to the other 3 issues, this one is the most important one for Mission, because any substantial reduction in size would threaten the success of the NGO's project plans (100).	Mission reports to Headquarters, and has to respect the laws of the host Government. Nevertheless, the services are vital to the region, and Mission has built up good informal relations with low level Government officials, the Churches and Village Heads (40).
Churches	Would on the one hand welcome an NGO of the same denomination to operate next to the local NGO of the other denomination, but also are aware that too much support might lead to troubling relationships with the Government. Considers a 20% decrease in size as acceptable (46).	The Churches are of the same denomination as Mission, but belong to the more conservative groups therein, whereas the NGO is known for its quite progressive methods regarding abortion, HIV/AIDS and therefore are reluctant to attach too much of a priority to this issue (50).	The Churches have good relationships and access to their members, who predominantly live in the most remote areas in this region. They can influence to what degree their members will be open to the services of the international NGO (60).
Government	The dominant faction of the ruling party prefers complete closure of the remote site for ideological reasons (0).	Realizes the important role the NGO plays for medical care in the region. Since it also anticipates resistance from Village Heads, this issue has not highest priority (60).	Can forcefully evict the NGO if necessary, and is therefore clearly the most powerful player (100).
Competitor	Since it is a local NGO of the same denomination as the Government, it has good connections to the Government. Would prefer to be the sole NGO in the area, but also benefits from CHH's presence, e.g. in logistics. Would therefore prefer a much smaller CHH unit (35).	Since it also to some degree benefits from CHH's presence, this NGO has no strong interest in CHH leaving (20).	Having entered the country later, informal relations to government are less well-developed, so informal power still is limited (20).
Villages	Health care in the region has significantly decreased since the refugee influx. They therefore prefer an expansion of the services by means of this international NGO, eventually to double its current capacity (140).	Collaboration with CHH and the related health improvements can significantly strengthen the position of the Village Heads in the population. Hence, this issue is relatively important to them (80).	Village Heads are subject to the national legislation, but given their strong moral influence on the population, they have considerable informal power (60).

Table 12.2 Summary of each stakeholder's position, salience, power for issue 1 ('size')

	Position	*Salience*	*Power*	*Effective power*	*Distance**
Mission	55	100	40	40	22
Churches	46	50	60	30	13
Government	1	60	100	60	32
Competitor	33	20	20	4	0
Villages	100	80	60	48	67

Notes

Values have been normalized (and rounded up to the whole number). The following example illustrates how to conduct this normalization, using the values for stakeholder 'position' on issue 1 as they are given in Table 12.1. Here, the numbers represent the preferred size of the remote unit in terms of staff (Mission: 70, Churches: 56, Government: 0; Competitor: 35; Villages 140). The following steps normalize the scale (which now ranges from a minimum of '0' to a maximum of '140' staff) to a scale from 0 to 100:

1 Determine the lowest value and label it 'A', and determine the highest value and label it 'B'.
2 For any value N between A and B in the original scale, calculate $X=(N-A)/(B-A)$.
3 The normalized result is $Z=100*X+1*(1-X)=90*X+10$.

For example: the normalized value for 'Mission' (preferred staff size: 70), is calculated as follows:

1 $A=0$, $B=140$.
2 $X=(70-0)/(140-0)=70/140=0.5$.
3 $Z=90*0.5+10=55$.

* Distance = the absolute distance between a stakeholders preferred position and the theoretical compromise position (see Step 4: outcome continuum).

course many different ways to determine a theoretical compromise. A good start may be to take the *weighted average* by summing up all stakeholders' positions, multiplied by their salience, divide it by the number of stakeholders and standardize the outcome value by dividing the result by 100. Once this is repeated for each issue, it may be useful to draw a graphical representation of the resulting outcome continuum for each issue by denoting the positions, the stakeholders representing each position and the compromise position. This graphical representation gives a useful first visual overview per issue.

Case illustration

Based on the data, the country management team calculates the theoretical compromise position. For the issue 'size', this yields 33.04. In detail, the calculation leading to this value looks as follows: ((Mission's position * Mission's salience = 55 * 100 = 5,500) + (Churches' position * Churches' salience = 46 * 50 = 2,300) + (Government's position * Government's salience = 1 * 60 = 60) + (Competitor's position * Competitor's salience = 33 * 20 = 660) + (Villages' position * Villages' salience = 100 * 80 = 8,000)) = 16,520/5 = 3,304/100 = 33.04.

Figure 12.1 gives a graphical representation of the outcome continuum with the compromise position. As can be seen, the theoretical compromise position happens to be identical with the position of the local NGO, i.e. the competitor. The other stakeholder who is relatively close to this outcome (13 points) is the association of

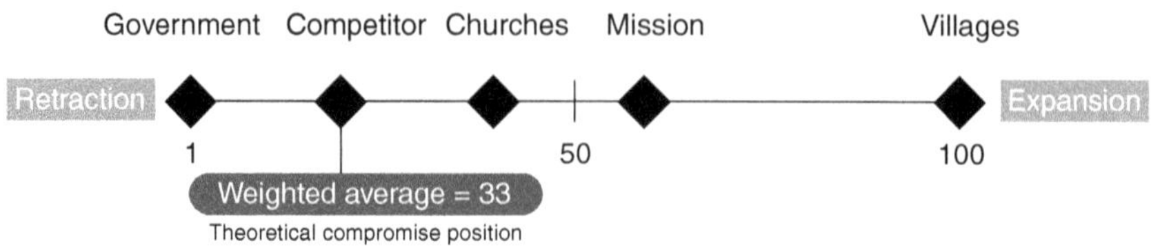

Figure 12.1 Graphical representation of the outcome continuum for issue 1 'size'.

Churches. However, this outcome is pretty far from Mission's position (22 points), and even farther from the position of the village heads (67 points).

Step 5: stability analysis

Main elements

The next step assesses to what degree the theoretical compromise would indeed represent a feasible – stable – outcome. This is done by analyzing how far from the theoretical compromise each stakeholder's preferred outcome on an issue is located and how much effective power these stakeholders have to actually block or support the theoretical compromise. The higher the number of stakeholders whose position has a high distance from the theoretical compromise position and the more effective power stakeholders have, the less likely it is that the latter represents a stable outcome, and the more likely it is that the stakeholders will engage in bargaining.

The visual tool for a stability analysis is a two-dimensional space per issue, in which the horizontal axis maps the product of salience and power of each stakeholder, and the vertical axis represents the absolute distance of each stakeholder's original position on this issue from the theoretical compromise position. The resulting two-dimensional space helps to classify stakeholders into four distinct categories, depending on which of the four cells they occupy:

1 Quadrant I (high distance from compromise, low effective power): these 'unhappy' stakeholders do not have enough effective power to influence the outcome.
2 Quadrant II (high distance from compromise, high effective power): these are stakeholders who are likely to challenge the outcome. They are powerful and have a strong interest in the issue.
3 Quadrant III (low distance from compromise, high effective power): these stakeholders have the incentive and the power to support the compromise outcome.
4 Quadrant IV (low distance from compromise, low effective power): these stakeholders either do not care enough or do not have enough power to exert active influence.

It is important to carefully choose the cut-off point on both dimensions. Though straightforward, simply dividing both dimensions at the 50 percent between the theoretical maximum and minimum of the scale (i.e. between zero and 100) may not always be the best option. This holds in particular for situations in which the distance between the stakeholder with the highest score and the stakeholder with the lowest score is relatively small, but where these small differences in position may still represent significant potential for disagreement or conflict. In such cases, choosing a 50 percent cut-off value on the absolute scale would cause most stakeholders to cluster into the same cell, though their positions would in fact conflict. To solve this problem, the analyst can choose a cut-off point in between the highest and lowest *realized* values. For example, if '80' represents the stakeholder with the highest position on an issue, and '50' represents the stakeholder with the lowest position on this issue, the cut-off value could be chosen half way between the two. This can be done by taking the absolute difference between the two values, divide it by two and add it to the lowest value. In our example this would yield 65 percent as the cut-off value (80 – 50 = 30/2 = 15 + 50 = 65).

Case illustration

The country management team draws the stability landscape by inserting each stakeholder on the respective coordinates for effective power (x-axis) and distance (y-axis). As cut-off points she chooses the 50 percent between the lowest and the highest score on each of these dimensions. For effective power, the highest value is 60 (for Government) and the lowest is 4 (for Competitor), see Table 12.2. This yields a cut-off point of (60 – 4)/2 = 28 + 4 = 32. For distance from compromise, we saw in step 4 that the theoretical compromise position is at 33.04. The stakeholder furthest away from the compromise is the government with 67 points. The stakeholder with the position closest to the theoretical compromise is the Competitor (33). Hence, the Competitor's distance from compromise is 0.4 points. The cut-off point is then (67 – 0.4)/2 = 33.3 + 0.4 = 33.7.

The graph (see Figure 12.2) immediately shows that Mission and Government fall in quadrant III, which means that they are likely to support the compromise outcome. Whereas they do not differ much in terms of their distance from the compromise outcome, there are considerable differences in effective power. For Mission this means that there might actually be room for a compromise with Government. However, the stability analysis also shows that the outcome is likely to be challenged by the stakeholder in quadrant II: the Villages. The graph shows that the Villages have both the interest and the power to affect negotiations on this issue. Finally, the Competitor and the Churches occupy quadrant IV: these stakeholders are close to the theoretical compromise position, but they have neither the incentive nor the power to affect the negotiations with regard to this issue. This implies that they have comparatively little influence over this issue. Note that the Competitor and the Churches, though both in this quadrant,

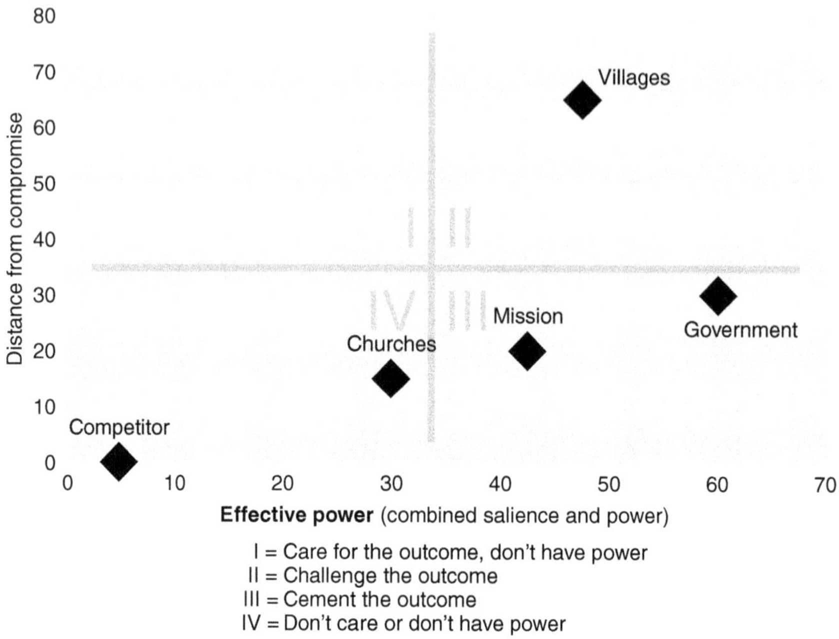

Figure 12.2 Stability analysis for issue 1 'size'.

differ considerably in effective power: the Churches are actually relatively similar to the Mission both in terms of distance from the theoretical compromise and effective power. Hence, Mission may eventually consider approaching the Churches for support at some stage in the negotiations. The position of the Churches in this figure also demonstrates that the classifications of stakeholders in such a figure should be interpreted with great care. In particular those stakeholders occupying a position close to the cut-off points deserve some closer scrutiny. The analyst should keep in mind that the classification is just an aid to support interpretation.

Step 6: stakeholder classification

Main elements

The major objective of the stakeholder classification step is to determine, for all stakeholders and across all issues, which stakeholders are allies, enemies or in-betweens for the focal stakeholder. A stakeholder is an ally (enemy) if the preferred position is close to (far from) the focal stakeholder's position. In order to classify a stakeholder into one of these three roles, it is necessary to define a cut-off criterion for the maximum distance within which a stakeholder still would 'qualify' as an ally. One possible procedure could be to first determine the distance to the most distant stakeholder, and then divide this figure by three in order to derive who is among the allies, enemies and in-betweens. Stakeholders falling in the first 33 percent would then be classified as allies, those in the second 33 percent would be in-betweens and those in the last 33 percent would qualify as enemies.

The resulting stakeholder classification provides two types of useful information. First, it shows the degree of likely support (or resistance) that the focal stakeholder may expect from each other stakeholder. For example, a stakeholder may be classified as an ally on all issues or only on some. Second, the stakeholder classification shows which issues are likely to be difficult or easy to realize. For example, the focal stakeholder may have no enemies on some issues, but many enemies on others.

Case illustration

The country management team now fills in the stakeholder classification table (see Table 12.3), using the position values from the stakeholder description step. For the first issue, it is evident that the Village Heads, though deviating from Mission's position in that they would prefer a larger unit, actually are allies, because Mission would not be against expanding the size of the unit. So Mission takes its own preferred position (55 on the normalized scale, as elaborated in Table 12.2) as the starting point for classifying the remaining stakeholders into enemies, in-betweens and allies. The category of enemies covers the first 33.33 per cent of the scale and therefore ranges from 0 to (33.33 * 55/100) = 18.3. There is one stakeholder in this range of the scale: the Government who has a position of 1 (see Table 12.2). The 'in-between' category ranges from 18.4 to (2 * 18.3) = 36.6 on the scale (i.e. the second 33.33 per cent). It contains one other stakeholder: the Competitor who has a position of 33 (see Table 12.2). And two stakeholders fall into the category of allies, covering the final 33.33 per cent of the scale (from 36.7 upwards): the Churches and Village Heads. The country management team repeats this operation for the remaining three issues (not elaborated in this example). The resulting picture table already allows some preliminary conclusions (see Table 12.3).

Table 12.3 Stakeholder classification (from Mission's perspective)

	Issue 1 size	*Issue 2 services*	*Issue 3 collaboration*	*Issue 4 timing*
Headquarters				
Churches				
Government				
Competitor				
Villages				

Note
White=enemy; grey=in-between; black=ally.

First, analyzing the columns, it shows that there is no single issue on which Mission does not have at least one enemy. But it also shows that there might be some room to maneuver: there is at least one ally on each issue, and three issues have also at least one in-between.

Second, analyzing the rows, the table shows that there is no stakeholder who would qualify as an ally across all issues – the Churches who are opposed to Mission's position on issue 3. Collaborating with the local NGO of the other denomination will not be supported by them. However, the table also demonstrates that the Government, who is the main enemy for the first two issues, actually is an ally for issue 3 and an in-between for issue 4. The strongest allies are the Village Heads, who share Mission's position for three issues, and are an in-between for one issue (issue 3). This stakeholder classification shows that none of the outcomes desired by Mission will be easy to realize.

Step 7: negotiation landscape

Main elements

The stakeholder classification is used as an input for the seventh step, in which the negotiation landscapes are mapped. The purpose of this step is to find out, for each issue separately, which negotiation strategy the focal stakeholder should use vis-à-vis each other stakeholder. Again, two dimensions are used to define the landscape. The horizontal axis represents power – the power of each stakeholder (see Table 12.2). The vertical axis represents salience (see Table 12.2) – the importance each stakeholder attaches to the issue under investigation. Based on the stakeholder classification carried out in the previous step, stakeholders are marked with a symbol or color identifying them either as an ally (black), an in-between (gray) or an enemy (white). All stakeholders – including the focal one – are then placed in this landscape, using their respective coordinates.

As during the stability analysis, also the negotiation landscape can be subdivided into four cells, for example by using a 50 percent cut-off value for both power and salience (for how to determine cut-off values, see the explanation in the section on Step 5: stability analysis):

1 Quadrant I (low power, high salience): these stakeholders have a strong interest in the issue, but lack the power and influence to have a strong impact on the negotiations concerning the issue. Allies falling into this quadrant are potential followers, and the focal stakeholder may want to bring them into the *coalition* if this can be achieved with comparatively little effort.
2 Quadrant II (high power, high salience): stakeholders falling into this category are labeled *shapers* because they are the ones who have both the interest and the power to have an impact on the negotiation. If there are

allies in this quadrant, attempting to develop a joint strategy with these allies may be a fruitful tactic for the focal stakeholder.

3 Quadrant III (high power, low salience): these stakeholders are potential *influencers*, because they have power, but they attach too little importance to the issue in order to translate their strong position into effective power. If there are allies in this quadrant, the focal stakeholder could gain some leverage by lobbying to increase their salience and support for this issue.

4 Quadrant IV (low power, low salience): since stakeholders in this quadrant are relatively powerless and disinterested with regard to the issue, not investing energy or ignoring these *bystanders* is the best option for the stakeholder. However, it should be noted that these stakeholders might nevertheless be very powerful and highly interested in one of the remaining issues. Hence, in order to determine which negotiation strategy should be used it is necessary to determine how a stakeholder relates to the focal stakeholder across all issues. This is done in the final step, the so-called relationship analysis.

Case illustration

The country management team now draws the negotiation landscape for each issue. Using the coordinates for power (x-axis) and salience (y-axis) as they were determined during the stakeholder description step, all stakeholders are placed on the landscape. To identify allies, in-betweens and enemies, the symbols and colors from the stakeholder classification step are used to denote each stakeholder. Again, the country management team determines the cut-off values by taking the midpoint between the highest and the lowest value in each of the two dimensions. Figure 12.3 shows the result for the first issue. Three stakeholders fall into the second quadrant, and two of them are allies (Churches and Village Heads). These 'shapers' have a high salience and comparatively much power. Developing a joint strategy with these two stakeholders therefore appears a viable option. The third stakeholder in this quadrant is the host Government. It has high power and the issue is relatively salient, but at the lower end of the quadrant. The graph further reveals that Mission itself is comparatively powerless with regard to this issue, though salience is high. She needs other stakeholders to support her. Occupying the fourth quadrant, the Competitor is a bystander who might be ignored. However, since the Competitor is an in-between and has low salience for this issue, Mission might consider to influence the Competitor in such a way that he attaches higher priority to this issue, and provide arguments that would lead the Competitor to endorse Mission's position. In case this succeeds, the Competitor would become a follower (quadrant I), who might add some extra weight to the coalition with the Village Heads and Churches.

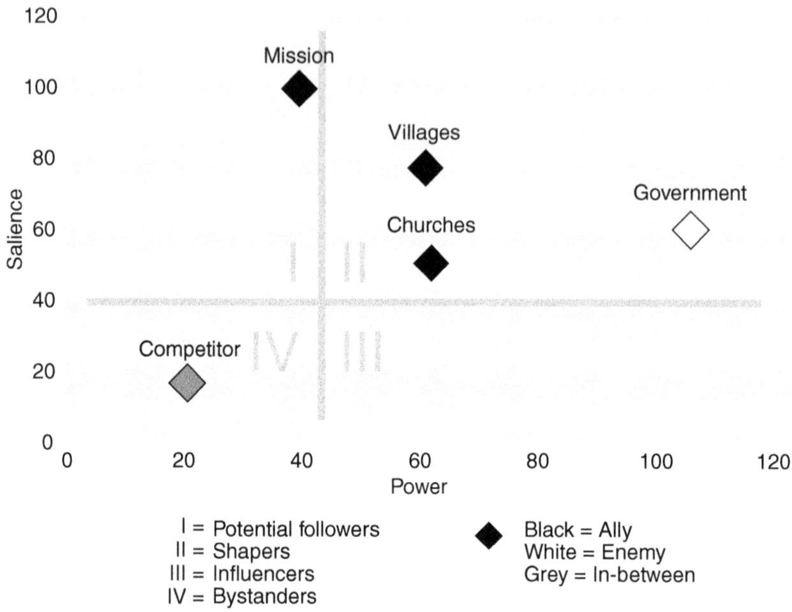

Figure 12.3 Negotiation landscape for issue 1 'size'.

Step 8: relationship analysis

Main elements

The objective of this step is to detect, for each stakeholder separately, potential bargaining opportunities, in which the focal stakeholder may exchange support for some issues of low importance to her for the other stakeholders' support on issues that do have a high salience for her. Again, a two-dimensional map is used to visualize this step. The horizontal axis represents the salience of the focal stakeholder; the vertical axis maps the salience of the specific stakeholder. Using the outcomes of the stakeholder classification step, symbols or colors are used to indicate whether the stakeholder is an ally, an in-between or an enemy with regard to each issue. Subsequently, each of the issues that are part of the stakeholder analysis are placed on their respective coordinates in one of the four quadrants:

1 High salience focal stakeholder, high salience other stakeholder: issues in this quadrant for which the other stakeholder has been classified as an enemy are potential *deal breakers*. Both have a strong interest in this issue, but their preferred outcome positions are relatively distant. Three negotiation strategies may be useful in this case (Allas and Georgiades 2001: 92): bargaining, lobbying and disaggregating (see also Bazerman and Neale 1994, for a more in-depth discussion of negotiation strategies). Bargaining

implies exchanging support for this or another issue for the other stakeholder's support on this or another issue. Lobbying implies attempts to lower the other stakeholder's salience concerning this issue, for example through interpersonal influence attempts. Disaggregating means to find possible subdomains for an issue: an opponent may be willing to make concessions on some of these sub-issues.

2 High salience focal stakeholder, low salience other stakeholder: issues in this quadrant can be *easy wins*, since the other stakeholder will invest comparatively little energy into their realization.

3 Low salience focal stakeholder, high salience other stakeholder: these issues are potential *bargaining chips* if the other stakeholder is an enemy. Since the other stakeholder has a much stronger interest in this issue, the focal stakeholder can offer to support her on this issue, in exchange for support on another issue.

4 Low salience focal stakeholder, low salience other stakeholder: these issues are of relatively low importance to both stakeholders and can therefore be ignored in the negotiation with this stakeholder.

Case illustration

In four separate graphs, the country management team maps Mission's salience against each other stakeholder's salience across all four issues. Colors or symbols mark whether the stakeholder is an enemy (white), an in-between (gray)

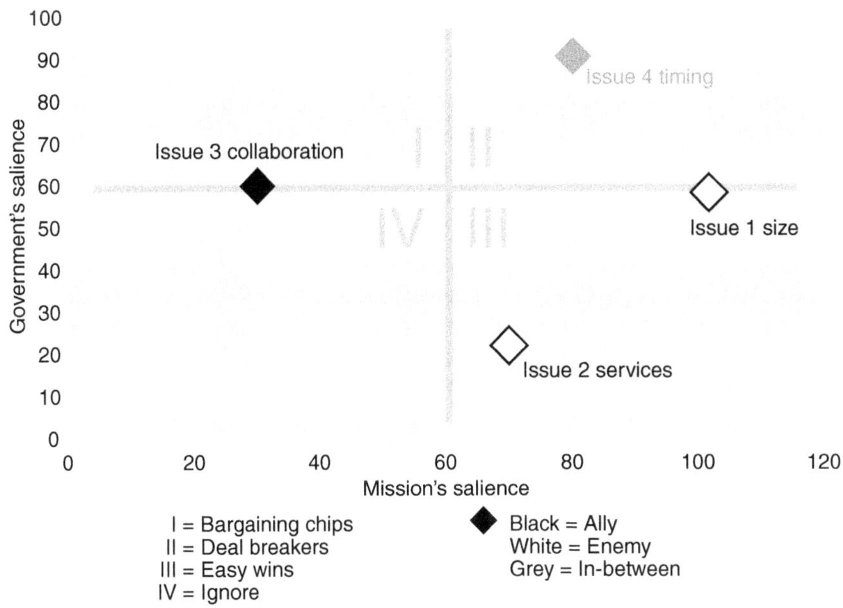

Figure 12.4 Relationship analysis for stakeholder 'government'.

or an ally (black) for a specific issue. Again, she divides the map into four quadrants for easier interpretation, using half of the distance between the lowest (20) and the highest (100) salience of both stakeholders as the cut-off value (i.e. 60). Figure 12.4 illustrates this map for the host Government. The map reveals the following.

First, since there is at least one issue in Quadrant I on which Government is an 'ally', Mission has at least one potential 'bargaining chip' that she might use in return for concessions on another issue: while Government attaches much value to Mission collaborating with the other NGO, Mission herself is relatively indifferent towards this issue.

Second, there are two issues in the second quadrant, which represents the difficult cases: size and timing. However, it also becomes evident that Government attaches much more importance to the 'timing' issue than to the 'size' issue. In addition, according to the stakeholder classification, Government is not an outright enemy on the timing issue but an in-between. Although initially against any outside interference, the Government now seems to be open to consider a short- or medium-term mission, but is not willing to allow a long-term mission.

Third, Government is an 'enemy' on the only issue in the third quadrant, 'services'. This issue is relatively important to Mission, but less so for the Government. The best thing to do for Mission is to keep this issue from the negotiation agenda as long as possible, and concentrate instead on the remaining three issues. More specifically, the map indicates that Mission most likely has to accept that serving the remote areas with mobile teams is not possible, but there seems to be considerable leverage with regard to when this has to be realized. Mission might opt for a bargaining strategy in which it offers to enter collaboration with the other NGO in return for an extended duration of its activities. This would meet the demands of the dominant coalition in Government – who does not want permanent or long-term presence of an international NGO due to ideological reasons – without them losing face, and it would allow the mission to carry on its work for a substantial period of time. Since the 'type of services' is relatively less important to Mission than is 'size', she may eventually also consider to only offer part of the identified suitable health activities with some of the services that the government approves of (e.g. on-site medical treatment in clinics and health centers) and not propose to conduct household visits, because the government fears losing or lacking control when these teams travel around.

As far as the 'size' issue is concerned, since both parties attach relatively much importance to it but find themselves on opposite ends on the position scale, a successful deal on this issue will be very difficult to achieve. If the negotiation strategy with regard to the remaining issues does not yield the desired results, Mission might therefore consider disaggregating the 'size' issue further, and build her propositions on the resulting sub-issues. For example, since Mission knows from the stakeholder description that the Government also would like Mission to concentrate on on-site medical care only ('services issue'), she may offer to appoint more local staff on these sites in order to reduce the visibility of international presence in the clinics.

In sum, this relationship analysis tells Mission that in her negotiations with Government, she should prioritize the timing of the activities, use the 'collaboration' issue as a bargaining chip first, and recur on the 'services' issue only if the outcome of this exchange is not yet satisfactory.

Conclusion

Technically, a stakeholder analysis is an instrument to map the relations of power and interest among a set of decision-makers. In practice, conducting a good stakeholder analysis is likely to encounter many challenges. The results stand and fall with how these challenges are solved. Two of these challenges are particularly important.

Quantification

Quantification – assigning numerical values to qualitative judgments about a stakeholder's power, salience and effective power – is one of the strengths of this approach, because it forces analysts to be explicit and transparent in the assumptions they make, and to be systematic in their reasoning. At the same time, quantification can be one of the tool's major potential pitfalls. First, numbers can evoke the illusion of precision. There will always remain some arbitrariness in the process of assigning numerical values to the evaluations. Yet, small 'errors' in the stakeholder description phase (say, salience of a stakeholder is estimated as '10' rather than '30') may lead to strongly diverging outcomes already during the stakeholder classification phase, and may ultimately result in contradicting negotiation landscapes. Replacing ranges of numerical values with seemingly less demanding 'qualitative' ordinal ratings (e.g. 'low, medium, high salience') will only partially resolve this problem. Therefore it is essential that the assigned values be carefully calibrated. Two steps may be useful for this purpose. First, instead of working with point estimates (i.e. a stakeholder's salience is '30'), the analyst assumes confidence intervals for each estimate (e.g. one stakeholder's salience is estimated to lie between, say 20 and 45, and another stakeholder's salience is estimated between 75 and 85). A downside of this approach is that it further complicates the analysis because it would require calculations carried out with the lower and the upper boundaries. The second option is to invest more effort into the reliability of the assessment, for example by attempting to approach additional experts or collect additional information on specific stakeholders. Of course none of these methods guarantees a 'correct' estimate.

Interpretation

Given the previous two limitations, it is evident that the results of a stakeholder analysis should be used as an instrument supporting a stakeholder in her interpretations that lead to strategy choices, rather than as a tool that delivers ready-made implementation strategies. As the case illustrations demonstrated, each case is different, and therefore requires careful consideration of the relevant context to avoid misinterpretations. For example, which options are feasible in a negotiation strongly depends on the general context, custom and rules in the setting – conditions that the stakeholder analysis can only indirectly account for in its stakeholder description phase. Hence, anything that goes beyond the dimensions covered and quantified in the stakeholder analysis is part of this context. Consequently, this tool can only be an aid in systematizing situations of complex interdependencies and power asymmetries.

Despite all these potential shortcomings, a systematic attempt to map the stakeholders and their estimated positions, salience and power forces the analyst to explicate his or her underlying assumptions. It shows which kind of information is needed, and therefore contributes to asking the right questions. And since it also can be easily visualized, it is a useful tool for both internal and external communication. Together, these elements make it a flexible and powerful instrument in any aid organizations' attempts to implement their missions in ever changing social and political contexts. Moreover, a stakeholder analysis helps aid organizations to make the step from suitable to feasible (and safe) interventions. In the fictitious case discussed in this chapter, we saw how the NGO started off with a set of three desired suitable interventions. Through the stakeholder analysis it became clear whether these three interventions were feasible given the stakeholder field. In addition, the stakeholder analysis resulted in relevant information as how to achieve the highest prioritized intervention(s) from the perspective of the NGO, by a strategy of negotiation with the government on issues of less importance to the NGO.

13 Monitoring, evaluation and learning in humanitarian organizations

Chamutal Afek-Eitam and Adriaan Ferf

Monitoring and evaluation as foundations for evidence-based humanitarian action

'Experience shows that the proper use of evidence may not prevail in decision-making in humanitarian action. Evidence-based decision-making encompasses external evidence, expertise and beneficiaries' values and circumstances' (Bradt 2009: 19). Throughout this book, it has been argued that evidence-based humanitarian action requires ongoing collection and analysis of adequate information which supports professionals in deciding whether a humanitarian intervention is appropriate or not, and in what capacity and mode it should be delivered. The production of such evidence can be focused on numerous topics such as the probability of disasters, their potential impacts, risks of escalation, the local, national and international needs and capacities to respond, the impacts of an intervention and the overall contextual situation in terms of political, social, economic, environmental, food, health and security matters.

To be able to analyze such information and to turn it into utilizable evidence for humanitarian organizations, meticulous strategic thinking and consideration is needed with concern to the quality of the information needed, its type and quantity, the systematic routines for its continuous collection and analysis, and the human resources and managerial support needed for these routines to take place. The task of collecting valid and reliable information from a multitude of sources and of transforming this information into utilizable evidence is thus undeniably challenging, particularly in always changing and fragile environments such as humanitarian crises (Frerks and Hilhorst 2002). This said, the production of such evidence is necessary to inform decision-making on what might work and what might not in certain situations, both at operational and strategic levels, in order to achieve effective and efficient programming and management. In the humanitarian assistance context investments in the generation and use of evidence are not only vital for organizational accountability and 'value for money', but also for moral reasons by ensuring that the best services possible are provided to those in need.

Monitoring and evaluation (M&E) mechanisms, as introduced in this (final) chapter, facilitate high quality programming by providing the information and analytical basis that is indispensable for evidence-based programming and decision-making in humanitarian action. Through M&E processes information is collected and analyzed before, during and after a disaster or crisis by means of a vast variety of methods. The rationale for an M&E framework and its corresponding processes is to attain higher quality management and service provision. The particular tools that may be applied in an M&E framework are diverse and depend largely on the particular agency carrying out the program or project. Yet, regardless of the tool selection and application, the principal product from all types of monitoring and evaluation activities would be the generation of reliable evidence, which would then feed back into the humanitarian program cycle and support evidence-based programming. M&E practices and mechanisms thus are a fourth crucial component in the H-AID framework as they provide the foundation for context, intervention and stakeholder analysis.

In the next section, we first discuss the evolution of monitoring and evaluation in the humanitarian sector and then proceed to define monitoring and evaluation and outline the fundamental purposes of M&E. Then sections follow on the place of M&E in the project cycle, as well as on various M&E types, tools and methodologies. We conclude the chapter with a reflection on the challenges that humanitarian organizations face regarding learning and accountability.

The evolution of monitoring and evaluation in humanitarian assistance

> Humanitarian work … is thought to be selfless, motivated by compassion, and by its very definition suggests good work … As relief is a gift, it is not expected that anyone (most especially the recipients) should examine the quality or quantity of what is given.
>
> (Harrell-Bond 1986 in ALNAP 2008: 1)

Until the mid-1990s, only few humanitarian originations recognized the obligation or even the importance of monitoring and evaluation approaches and methods as mechanisms to ensure quality of services or appropriate management of aid. Nonetheless at the same time in the development sector monitoring and evaluation were already a standard practice. Compared to the development sector, in the humanitarian sector external and internal requirements for organized thought were significantly less sought after. This lack of interest by the humanitarian sector is explained by the following reasons (ALNAP 2008).

First, externally, relief work was very much seen by the general public, agencies and donors, as 'doing good' by selfless hard working people, motivated by compassion, who were prepared to go and help wherever whenever needed. This interpretation resonates the idea of delivering 'a gift' of expertise, supplies or money. Fitting with this public image, there were no external pressures (from donors, for example) to examine whether humanitarian work was of good quality, or relevant, effective and efficient.

Second, humanitarian aid pertains to life saving and relief operations where speed and urgency of response are of utmost importance. With this, the sector predominantly consisted of individuals who could operate and deliver aid under challenging circumstances. Consequently, the sector lacked more process-driven and reflective individuals. Third, during this period, baseline data was scarce or absent due to lacking operational institutions and appropriate methodologies for information gathering and analysis that could address the scarcity of information and the often-short project cycles of humanitarian response. Fourth, as staff turnover was high and information systems were nearly nonexistent, institutional memory was inadequate leaving little room for improvement. The sector thus had a strong action-oriented culture with little room for systematic reflection and learning. Consequently, the costs of setting up and running monitoring and evaluation systems were deemed to be high and the sector avoided investing in M&E. This overall attitude halted documentation of process tracing, learning, monitoring and evaluations. The limited possibilities to generate information had consequences for the affected areas, communities and the operating organizations and regularly resulted in inadequate programming.

A great deal has changed since 1986. Progressively, the past three decades reflected an increased demand for greater monitoring, evaluation and accountability of INGOs (Ebrahim 2005). This demand was driven by a variety of developments and factors. A prominent factor was a rapid growth of the number of NGOs worldwide (Smith 1993, Wolch 1990), another is the exceptional number of wide-ranging humanitarian interventions that took place in Rwanda, Somalia, former Yugoslavia, Kosovo, southern Africa, East Timor and Central America in the 1990s and beyond. Furthermore, there were a few highly exposed scandals, which dramatically affected and weakened public and media confidence in INGOs (Gibelman and Gelman 2001), such as the Rwanda genocide in 1994 and the sex-for-aid scandals in West Africa (HAP 2008).

Alongside these developments, a discourse regarding the lack of learning developed which led to increased pressures on humanitarian agencies for greater transparency and accountability (Frerks and Hilhorst 2002). These debates involved practitioners, donors, research bodies, UN agencies, INGOs and academics surrounding humanitarian assistance (GHA 2007/2008). Consequently, governments heightened their funding requirements, and with these their reporting and accountability requirements, forcing the humanitarian sector to seek information and evidence. The sector thus started to invest in the design and institutionalization of internal and external quality assurance practices, monitoring and evaluation included.

The efforts to address these quality assurance challenges were also reflected in a large number of international conferences (such as conferences on food security in Rome in 1996 and on climate change in Kyoto 1997), next to reports and initiatives to develop quality standards for the sector, and tangibly manifested themselves through a variety of monitoring and evaluations approaches. Examples of such are the Active Learning Network for Accountability and Performance founded in 1997 (ALNAP), the Humanitarian

Ombudsman Project (founded in 1998), the Humanitarian Accountability Project (HAP), the accountability initiative of the Emergency Capacity Building project (ECB), the Sphere Project for technical aspects of disaster response and the DAC-OECD criteria which defined standards for monitoring and evaluations (i.e. the OECD humanitarian evaluation criteria). The latter initiative launched the now commonly shared performance standards of relevance/appropriateness, connectedness, coherence, coverage, efficiency, effectiveness, impact and were considered a pioneering effort to improve effectiveness and enhance accountability in the sector (see Buchanan-Smith and Cosgrave 2013 for the definition of each criterion). It laid the foundation for quality assurance practices in humanitarian assistance and promoted debates inquiring what comprises and signifies humanitarian aid (ALNAP 2001). Nowadays, there is common understanding that the number of monitoring and evaluations efforts has grown ever since (Clarke and Ramalingam 2008, Hallam 2011) and that with this a heightened professionalization of the humanitarian sector was achieved.

However, some criticism remained. While monitoring and evaluation frameworks have focused mostly on outputs, outcomes and processes, they provided little evidence for the impacts of humanitarian operations. The legitimacy of humanitarian assistance is therefore still repeatedly – publicly and internally – challenged, whereas at the same time the demand for high quality aid continues to intensify (Ebrahim 2005, Carroll and Phillips 2010, Levine *et al.* 2011, Binder 2013). Such scrutiny, in combination with increased interest from the media and the general public, continues to keep pressure on humanitarian and donor agencies to improve their impact and accountability.

Nowadays, the attention in humanitarian evaluations and monitoring exercises is focused on inquiring the impact of humanitarian assistance in order to be able to appreciate and measure, in an evidence-based way, how aid affects the lives and livelihoods of its recipients (ALNAP 2008). In order to investigate these matters, the importance of involving recipients has been stressed through methods such as impact assessments, monitoring practices and robust needs assessments. Humanitarian organizations are through these methods able to generate evidence of what type of assistance is most appropriate and efficient (Proudlock *et al.* 2009): 'It is now slowly becoming the norm for "beneficiaries" to be involved at some stage of the evaluation process, although this is usually only as informants rather than setting the evaluation agenda' (Hallam 2011: 15).

Monitoring and evaluation: definitions and purposes

The systematic collection of information is essential for both monitoring and evaluation. The two are closely interconnected, yet do not fully depend upon each other. The two differ in their actual setup, function and impact. Ideally, the two inform each other but this is not always the case (Perrin 2012). Monitoring is defined as:

> A continuing function that uses systematic collection of data on specified indicators to provide management and the main stakeholders of an ongoing development intervention with indications of the extent of progress and achievement of objectives and progress in the use of allocated funds.
>
> (OECD/DAC 2002: 27).

Evaluation is defined as 'The systematic and objective assessment of a planned, on-going or completed project, program or policy, its design, implementation and results' (OECD/DAC 2002: 21).

In short, monitoring focuses on the collection of information. The type of information to be collected should be based upon specified predefined indicators outlined in the project and program planning stages. These indicators help to determine the extent of progress made towards achieving the project or program goals and objectives. Evaluation investigates whether the objectives of an ongoing or completed project, program or policy have been achieved or had the expected impact. Evaluation aims at providing credible and useful information and enables the incorporation of lessons learned in ongoing or future projects, programs or policies.

Monitoring is mostly conducted internally and occurs routinely throughout projects or programs during the complete project cycle. An organization will conduct monitoring for different reasons: to ensure whether a project is on track with attaining its goals (e.g. which share of the target group is getting drinking water or restored their income sources), to keep track of changes in the context to assess whether the planned intervention is still appropriate (e.g. whether the previously identified target group still needs support or needs a different type of support) or to keep an eye on the security situation (e.g. to assess whether other areas should be considered or if it is still safe for the organization to operate in the context).

Defining quality performance and outcome-based indicators is a critical aspect for an effective monitoring and evaluation plan. Effective monitoring implies focusing on indicators on all aspects and stages of the project including input indicators (resources: time, money, material and staff), process indicators (leading from resources to results/outputs), action/output indicators (leading from action to results/outputs), outcome indicators (leading from results to purpose/outcome) and impact indicators (leading from purpose to the attainment of the overall objective). Elements dealing with outcome and impact indicators obviously fall into the category of evaluation (Moneval Solutions 2013).

While different parts of the organization may have different monitoring responsibilities for different reasons, the data gathered by all should be compiled and organized in such a manner that it can be sorted and made sense of for purposes of analysis. From such an analysis, it should be made possible to generate valid and reliable findings as input for management decisions that are evidence based. As monitoring is supposed to be routinely conducted, 'good monitoring' thus enables constant quality assurance of operations, enabling – if needed – room for instantaneous adaptations to project or program activities.

Evaluation is conducted at specifically preplanned and clearly defined moments. Good evaluation acts as an objective assessment of the intended and unintended project achievements and impacts. The focus of the evaluation may be decided upon already during the initiation of a program (in the proposal) or might be based upon an emerging need during a project's implementation.

Monitoring and evaluation processes are different but interconnected processes. In well-designed monitoring and evaluation frameworks the information needs of the evaluation are taken into account in the selection of the monitoring indicators. This enables the evaluation to build upon longitudinal data sets on relevant outputs and outcomes and draw conclusions that have a strong evidence base.

Both monitoring and evaluation may have a number of purposes. Below some principal purposes of both monitoring and evaluation are mentioned (cf. Frerks and Hilhorst 2002, Hallam 2011):

- strategic: to initiate, continue or terminate an operation and to decide on timing, approaches and resources of a humanitarian operation;
- program design and operations: informing and supporting programming, implementation and possible modifications to programs;
- organizational learning at project, program, organization or sector level;
- accountability: the process of taking account of and being held accountable by different stakeholders and primarily those who are affected by the exercise of power (HAP 2008);
- impact measurement.

Monitoring and evaluations often aim at serving more than one purpose; M&E collects and analyzes data for operational decision-making as well as for accountability to donors and/or recipients, next to learning. However, as the different purpose have often different approaches and operative requirements, research has shown that it is advisable for M&E frameworks to select a well-defined single leading purpose (Hallam 2011). This will help ease and streamline the processes of planning and implementation and increase efficiency and effectiveness of the M&E activities.

Depending on the purpose, the implementing organization(s), and the intended audience of the monitoring or evaluation exercise, different forms, approaches and methods of M&E can be used, as we will outline later in this chapter. But first, we identify the place of monitoring and evaluation in the project/program cycle.

Monitoring and evaluation and its place in the project/program cycle

Monitoring and Evaluation are part of an ongoing data collection process; it is a sorting and analytical process conducted throughout the whole project and program management cycle (see also Chapter 1). The core elements, which are generally present in all management cycles, include: *Identification*, *Formulation*, *Implementation* and *Evaluation* (see Figure 13.1).

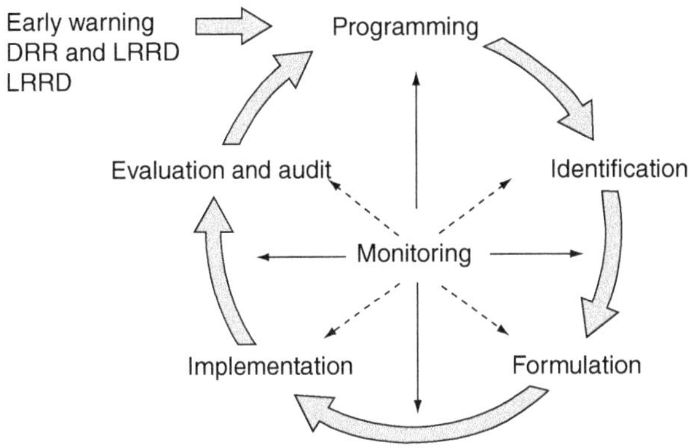

Figure 13.1 Data collection and analysis cycles used in the project management.

More specific elements – that often differ between organizations – may appear under each of the previously mentioned core elements of the cycle. Some examples of these are early warning mechanisms, needs assessments, real-time evaluations and impact assessments. Each element includes particular requirements for data gathering (for example, on gender, age, capacities and participation), analytical methods to apply (such as actor analysis, problem trees, SWOT or Intervention Analysis) or particular types of planning and programming tools (for example, log frames and GANTT charts). Although each element has its own specific purpose, they are all closely linked and their effectiveness depends on one another, therefore the circular shape.

In many program and project cycle visualizations, the monitoring and evaluation elements are either placed in the middle of the cycle, on an outer sphere surrounding the cycle or at the end of the cycle with arrows pointing towards every other stage of the cycle. Whatever the place of M&E in these visualizations, the purpose is to signify that monitoring and evaluation activities are ongoing activities throughout the lifetime of a program or project. Ongoing monitoring and evaluations activities would essentially mean that programs and projects repeat the cycle numerous times and, more than often, run back and forth, as newly generated information on the status of the program or the project brings about reassessment and the need to redesign.

M&E types according to project phases, purpose and implementation modalities

In the previous section several evaluation and monitoring types related to the different phases of the project cycle have already been mentioned. It shows that evaluation and monitoring covers a wide range of types that can be categorized according to phases of the project cycle, as well as to its purpose, implementation modalities and content (see Table 13.1).

Table 13.1 Overview of monitoring and evaluation types and modalities

	Definition	*Focus*	*When*	*Why*	*Who*
Early warning	A system to forecast and signal hazard or conflicts	Specific risk in geographical area	*Ex ante*	To prepare for timely action	International and national institutions as well as communities
Needs assessments	Providing a rapid overview of the emergency situation: mapping the immediate impact of the crisis, estimating direct needs of affected population and defining priorities for humanitarian action	Needs of affected groups and communities	*Ex ante*	To identify people in need, their vulnerabilities and capacities, and needed support in order to design adequate projects	Humanitarian organizations
Baselines	An analysis describing the situation prior to an intervention, against which progress can be assessed or comparisons made	Condition of affected groups or communities prior to intervention	*Ex ante*	To provide a basis to assess progress and results against in order to make comparisons	Project/program staff External teams
Performance and context monitoring (formative)	The tracking of project outputs and outcomes as indicators of project effectiveness, or the extent to which the project achieves its stated objectives (performance) The tracking of contextual changes and trends related indicators that impact upon project relevance, performance and operations (context)	Project/program indicators Thematic Geographical Process	Throughout the project cycle	To ensure information provision to various levels of management and external stakeholders for timely and well-informed decision-making To provide information to evaluative processes	Project/program staff (Head) office staff In participative monitoring recipients participate in data collection and analysis
Real time and mid-term evaluations (formative)	Evaluation performed at an early stage (real time) or towards the middle of the period (mid-term) of implementation of the intervention	Project progress, strategies and process and context	Early phase of the response and mid-term	To provide a distant review of relevance and performance to timely improve operations and to adapt (if needed) programming	(Head) office staff, external experts

After action review Appreciative inquiry (formative)	*After action review*: a structured review or debriefing process for analyzing what happened, why it happened and how it can be done better *Appreciative inquiry:* a solution-oriented approach that asks positive questions around what works	Performance or strategies of specific project/program interventions	During or after the project	To learn from practice to improve performance and strategies	Staff of one or more projects/ programs
Summative or formative final evaluations	Assesses the extent to which a project/program achieved intended outcomes and the delivery processes	Project outcomes and processes	End or just after end of project/ program	Summative final evaluation: to account to management levels, donors and stakeholders Formative final evaluation: learn from strength and weaknesses	Summative evaluations: external by experts formative evaluations internal or external experts In participative evaluations recipients participate in data collection and analysis
Impact assessment	Impact evaluations measure whether the intended goals are being achieved or not and why and how the expected impact was achieved or not; and whether the intervention had impact (or not)	Intended and non-intended impacts of operation	At the end of after project/ program	To establish which strategies have worked and had impact	External experts
Meta-review	Study the effectiveness of a number of projects/programs of one or more organizations	Geographical areas Sectors Themes	After end project/ programs	To establish which strategies and/or processes were applied and effective	External

Evaluative mechanisms in different phases of a project or program

Some M&E types (see Table 13.1) can be applied *ex ante* (before) the project start, such as early warning mechanisms, needs assessments and baseline studies. During project or program implementation, performance and context are ideally monitored and for this real time, mid-term and/or after action reviews and appreciative enquiries can be used. At the end of the project cycle a final evaluation or an impact assessment can be conducted. Impact can also be assessed several years after the project ended, a time at which normally meta-evaluations take place. For example, Ian Christoplos *et al.* (2009) conducted an impact assessment ten years after Hurricane Mitch struck middle America.

Meta-evaluations assess the findings of a – sometimes very large – number of evaluations simultaneously. This can be done by individual organizations, for example to assess the effectiveness of all their evaluated projects with respect to gender or private sector involvement, or for all evaluations available on a certain theme, such as to compare the effectiveness of cash transfers to direct food distributions. By meta-evaluation approaches the full value of the many individual and costly expensive evaluation exercises can be extracted (Hallam 2011).

Evaluative mechanisms and their relation to the M&E purposes

Evaluative mechanisms can also be categorized according to their purpose. For example, formative monitoring and evaluations are first and second loop learning oriented and intend to improve performance. Good examples of formative evaluative mechanisms are real time evaluations and appreciative enquiries. Real time evaluations are undertaken at an early phase of the response when key operational and policy decisions are being taken. They can help to identify important issues that have been overlooked in the heat of the emergency response (Buchanan-Smith and Cosgrave 2013).

Appreciative enquiries ask questions to identify what works, adopting a constructive approach by being solution oriented rather than problem solving. While formative mechanisms focus on improving operations and learning, summative evaluative mechanisms determine the extent to which the anticipated outputs, outcomes or impacts were realized and provide information about the worth of a project/program.

Evaluative mechanisms and implementation modalities

Monitoring and evaluation exercises can be conducted by the team or organization that implements the project or program. This is called internal or self-monitoring and evaluation. It can also be done externally by staff of organizations operating in the same area or sector (also called a peer review), by independent external evaluators or by mixed teams. Involvement of staff of the implementing organizations in evaluative mechanisms is argued to enhance access to information and learning while external or independent monitoring and evaluation is considered to be more objective and a condition for accountability (Hallam 2011).

There is a general understanding that direct stakeholders and direct beneficiaries should always be consulted in any monitoring and evaluation exercise. In participative monitoring and evaluation their role is considerably more important. In this type of M&E activity these actors share responsibility of the monitoring and evaluation results. Representatives of agencies and stakeholders (including beneficiaries) are also involved in the design, implementation and interpretation of the findings. Participative evaluative practices aim at enhancing ownership and utilization, to diminish evaluation biases by joint critical reflection, and to empower stakeholders through increasing their knowledge and understanding.

Also, humanitarian organizations or donors can work together in monitoring and evaluation (joint monitoring and evaluation). A well-known example of a joint evaluation is the Tsunami evaluation by the Tsunami Evaluation Coalition, a collaborative evaluation effort of over 40 humanitarian and development cooperation agencies after the Indian Ocean Tsunami in 2004 (TEC 2006).

While most evaluative mechanisms relate to individual projects or programs, they also can relate to a set of interconnected activities, projects or programs (cluster), a specific theme (such as peacebuilding and conflict prevention or a cross-cutting theme such as gender or children), a process (for example, complaints mechanisms), a sector (related to water and sanitation or food security interventions) or a specific region or country program of one donor or international organization (such as the Great Lakes program of the EU-ECHO).

The different forms of evaluative mechanisms are not mutually exclusive. On the contrary, they often overlap: for example, an external summative evaluation might well provide important lessons learned to improve performance (formative), be organized by several agencies together (joint) and covering a theme (thematic) in a group of programs (cluster). Table 13.1 gives an overview of the different forms.

Monitoring and evaluation methodologies and tools

M&E guides of the main humanitarian agencies[1] provide a wide range of approaches, methodologies and tools to monitor and evaluate project/programs. International conventions, the Code of Conduct for the International Red Cross and Red Crescent Movement and NGOs in disasters, and the HAP Accountability Framework set the values and principles. The OECD-DAC defined the evaluation criteria and the Sphere Project the standards for technical quality. For specific sectors such as food security or livestock recovery and themes like gender, there are specific quality standards.

Goals-based, goals-free and theory-based M&E approaches

Monitoring and evaluation approaches are generally a combination of goals-based, goals-free and theory-based approaches. Goals-based approaches focus especially on intended outputs, outcomes and impacts, and are important for accountability, but might not identify and examine unanticipated results. Goal-free approaches identify all outcomes and impacts, whether anticipated and intended or not. Their strength is the identification of unanticipated results – which is common and important in humanitarian contexts. However as donors are largely interested in the delivery of the pre-intervention determined results (i.e. intended impact and outcomes only), they are less suitable for accountability purposes regarding donors.

Theory-based approaches, as also discussed in Chapter 11, are guided by a project or program theory, defined as the logical relation between all parts of a project, as provided by the logical framework or by the evaluation team as (re) constructed in a theory of change. Both logical frameworks and theories of change explain how program activities and results are connected with each other and contribute to achieve results at different levels. They provide a better understanding of the fundamental mechanisms underlying a project or program and facilitate understanding of what works, when, where, why and for whom. These approaches are however time consuming and can be biased as the logic might not cover all important outcomes and aspects (Sasketchewan 2008).

Quantitative and qualitative methods in M&E

The monitoring and evaluation methodology further defines the key questions to be answered (e.g. concerning output, outcomes, impacts, related processes), the indicators for each question, the source where data can be found (e.g. statistics, documentation, records, stakeholders and other actors) and the benchmark or targets set for the assessment. The questions, indicators, sources and benchmarks can be combined in a monitoring and/or evaluation grid.

Which tools to use to collect data on the predefined indicators depends in principle on the evaluative approach and the type of monitoring and evaluation questions that need to be answered (see also Table 13.2). Quantitative methods provide uniform measures of project progress, outcomes and impacts, for example on the number of people provided with clean drinking water or the decrease in drinking water related diseases ('who', 'how much' and 'how' questions). Good quality statistics or surveys with representative sample sizes ensure that the findings can be generalized to a wider population. Statistical analysis also enables establishing causalities between outcomes and impacts. Quantitative methods are often considered as more robust and objective and therefore provide more convincing evidence for the effectiveness and impact of strategies and are thus more appropriate for learning and accountability purposes. Examples of quantitative tools are face-to-face, mail or telephone surveys (such as nutrition surveys), process tracking records, direct measurement of indicators (e.g. blood pressure) and service utilization records.

In order to explain observed outcomes and impacts of a project or program (the 'how' and 'why' questions) quantitative methods are however less appropriate. Furthermore, they require specific skills of the staff and are often more expensive and time consuming than qualitative methods. Qualitative methods

Table 13.2 Qualitative and quantitative applied data collection methods

	Early warning	*Needs assessments*	*Monitoring*	*RTE/MTR*	*Final evaluations*	*Impact assessments*	*Meta-evaluations*
Quantitative and/or qualitative data collection and analysis?	Largely quantitative, complemented with qualitative data, e.g. through community monitoring mechanisms	Largely qualitative especially in the relief phase, later increasingly quantitative	Largely quantitative, input–output information, rarely completed with qualitative data on context, target groups and effects	Largely qualitative	Largely qualitative	Largely quantitative	Mixed, depending on the information available in the evaluation reports
Types of data collection methods used	Trends based data collection on incidents, weather, livelihoods, harvests, market, income risk analysis and scenario planning	Context analysis Individual and group interviews, focus groups and direct observations Questionnaires and surveys	Progress tracking on indicators financial data Surveys sometimes complemented with interviews, focus groups, observation and community monitoring Rarely outcome data collected with e.g. most significant change method	Document review Monitoring data Interviews/focus groups with recipients, key informants and stakeholders	Document review Interviews/focus-group discussions with key informants and stakeholder Direct observation Story telling	Surveys, quasi-experimental methods, Rarely random control trials	Document review Statistical analysis methods

are more flexible, allow for open-end questions, can explore a topic more thoroughly and are therefore more appropriate to collect data to answer these questions. Qualitative methods are also relevant to understand backgrounds and identify topics and questions for later quantitative data collection. Examples of qualitative tools are open or semi-structured individual and group interviews, focus group discussions, workshops and meetings, story-telling techniques such as the most significant change method and direct observation.[2]

The literature suggests that not one approach is best, as each has its own particular set of strengths and limitations. Using a mix of both different quantitative and qualitative methods for monitoring and evaluation is generally felt as most appropriate as they complement each other's strengths and will deliver more coherent, reliable and useful information (Adato 2001, European Commission 2007, ALNAP 2013). Using different methods also allows for triangulation of information and obtaining reliable evidence through verification of data by comparing results of different methods and avoiding bias. In practice needs assessments and evaluations favor largely qualitative above a mix of quantitative and qualitative methods. For monitoring, project management has the tendency to limit monitoring information to quantitative data collection, with limited application of qualitative methods.

Other methodological considerations play a significant role in the choice of data collection methods. In the early phases of a humanitarian crisis, speed is an important factor, while the operations normally take place in data-poor, politicized and complex environments in which physical access is limited and populations are mobile. Furthermore, financial and human resources for monitoring and evaluation are often limited in the initial phases.

Challenges to learning and accountability in humanitarian organizations

Monitoring and evaluation play a key role in humanitarian organizations' requirements and endeavors to learn and improve the quality of their work. As mentioned earlier, the main purposes of monitoring and evaluation are generally towards facilitating and improving accountability, operational matters and learning about the impact of projects and programs as well as on the organizational level. However, the question of how to achieve these aims of learning and accountability is not easy to answer. What would be a complete and effective strategic framework for M&E to attain these purposes in humanitarian organizations? Drawing from organizational theory on learning we will present a framework, which may provide a start in answering the above question and assist in the complete, efficient and effective operationalization and institutionalization of M&E in humanitarian organizations.

The 4i framework for organizational learning

Marie Crossan designed the 4i framework in order to outline a wholesome organizational process to attain organizational learning (Crossan *et al.* 1999). We have adapted the 4i framework for our needs in order to shape a thorough framework for strategically managing and reaching the variety of purposes of different M&E mechanisms (Afek-Eitam forthcoming). Crossan's framework divides the process of organizational learning into the 4i's: Intuiting, Interpreting, Integrating and Institutionalizing.

In the humanitarian sector, the uptake of any evaluative or monitoring activity involves the first process of *Intuition*, driven by an individual who expresses the need to conduct monitoring, an assessment or an evaluation. The *Interpretation* process then follows this, as the group considers the individual's notion. Once the group recognizes the individual's notion to start an M&E process, *Integration* processes start taking place. Especially in the humanitarian sector these processes are challenging to fully achieve. *Integration* processes occur amongst teams and groups developing and agreeing on action. In this process there is a risk that plans are made quickly and do not draw on evidence, thus making evidence-based decisions at this stage weak(er). For example, terms of references of evaluations might not draw upon past evaluation findings. Last, in the crucial *Institutionalization* process, also referred to as the systematic uptake of findings and decisions, there is a risk that findings are not disseminated or fed back into the organizational routines and systems.

According to Crossan these processes are crucial stepping-stones towards organizational learning. Each stepping-stone is owned by three different learning levels of an organization: Intuiting and Interpreting take place at the individual level, Interpreting and Integrating take place at the group level with the Interpretation process overlapping with the individual level, and Institutionalizing takes place at the organization level.

We find Crossan's framework valuable as it helps to define the necessary stages to attain organizational learning by identifying specific owners and particular actions they need to take. Highlighting the particular organizational learning owner level is particularly relevant and useful to the humanitarian arena as there is risk that this ownership can be lacking at the end of an evaluative process. Pointing these out in the framework and the various strategic stages helps insure that as the planning process evolves more thought and commitment is spent on 'who owns what: action-decision-action' issues during an evaluation mechanism. This then leads to the smoother planning and implementation of another often-lacking process: follow-up. Furthermore, the 4i framework helps in unfolding and operationalizing the typical widely used broad concept of Planning, which is often open to many individual and organizational interpretations and applications, resulting in the risk of deficiencies in its application. Crossan does this via coining the *Interpretation* and *Integration* processes.

For our aim, to outline a strategic framework which is tangible and efficient for implementing effective evaluative mechanisms in the humanitarian arena, we altered Crossan's i's, see Figure 13.2. We added an additional Individual/Group level process that may occur before or after the *Integration* stage: *Implementation* (Afek-Eitam forthcoming). *Implementation* processes regularly take place through data gathering, analysis and reporting and can be a stepping-stone to institutionalization.

Level	Process	Inputs/outcomes
Individual	Intuiting	Experiences Images Metaphors Obligations
	Interpretations	Shared experiences Language Cognitive map Conversation/dialogue
Group	Implementation	Use of capacities (individual, group, organization) Use of systems (organization) Record of process/experiences
	Integration	Shared/formal understandings Mutual adjustments and commitments Agreed interactive systems Use of record (evidence/experience)
Organization	Institutionalization	Routines Dissemination Diagnostic systems Rules and procedures

Figure 13.2 5i framework for implementing evaluative practices (source: based on Afek-Eitam forthcoming).

To illustrate this for the humanitarian sector: When a particular topic that came up during the *Interpretation* process has sufficient prior evidence, the group may choose to *integrate* these findings and *implement* their activities accordingly, making an evidence-based decision which will later, depending on its outcome, be *institutionalized* in one manner or another. An example for such an *Integration > Implementation* flow would be the preparation of an evaluation Terms of Reference[3] and the implementation of an evaluation. The outcome of the implementation process (e.g. an evaluation report) would then be institutionalized. What would normally be called Planning is divided here into three different processes: *Interpretation, Implementation* and *Integration*, each containing different specific planning inputs and outcomes elements as well as other non-planning features. See Figure 13.2 (Afek-Eitam forthcoming).

The structure of Figure 13.2 seems to be linear, however it is key to note that the process of implementing an evaluative mechanism is a spiral one, starting at the individual level and ending at the organizational level. Further, the flow of the framework is dynamic; until reaching the organizational level processes, the processes and the actions conducted by either the individual or group levels might go up and down the spiral numerous times until sufficiently justified progression has taken place to be taken into the organization level processes. The spiral is then repeated within an evaluative mechanism until completed. It is then repeated between them, utilizing and building upon past knowledge.

The 5i framework and its challenges for humanitarian organizations

Failure to meet the challenges related to the above mentioned processes might obstruct an M&E activity from attaining its purpose. If the chain of processes is broken, it can be expected, based on the premises of the 5i framework, that the result will be a lack of utilization of evaluations next to risks of repeated mistakes in the delivery of aid and a lack of evidence that can support the building of actual solutions (Afek-Eitam forthcoming).

Inefficient practices of information management bring about what Argyris and Schön (1978) would call single loop and double loop learning and disable triple loop learning (Argyris and Schön 1974). Single loop and double loop learning result in adjusting to undesirable situations or outcomes without understanding or treating the root causes of the faulty action. Consequently, the learning process is not retained and the same occurrence is likely to repeat. In order to empower and attain triple loop learning in the humanitarian sector, by which the root causes of faulty action can be identified, data should be entered into some sort of institutionalized system that would allow analysis, tracking of patterns and evidence generation which are all needed for deeper transformative change. For example, this would be possible during an evaluation study which studies both the organizational conditions that might be required to achieve better programming as well as decisions concerning wider policies, programs or projects.

Put differently, if monitoring information is not collected, poorly collected, wrong information is collected or when there is no documentation, an evaluation mechanism will generate little data to draw on, making it less rigorous, valid and reliable. And, once evaluation findings are not institutionalized and disseminated, managers who are planning new programs cannot draw on past organizational experiences, leading them to possibly repeat mistakes or conceive faulty or not optimal practices and accept faulty monitoring practices, ending with deficient monitoring and evaluation mechanisms and a broken M&E cycle. The 5i framework (Afek-Eitam forthcoming) outlines processes that are expected to prevent such deficiencies from happening as well as the challenges that humanitarian organizations face in their efforts to utilize the full potential of M&E activities and routines.

Conclusion

Monitoring and evaluation are important to provide evidence to be used in programming. This chapter argues that much progress has been made in the last two decades. However, it proves to be challenging to organize and institutionalize M&E practices optimally in such a way that it helps to improve operational performance and programming.

Evaluative mechanisms and practices are clearly not (yet) the silver bullet that will deliver all evidence for evidence-based programming. However, the regular analysis of evaluative practices by ALNAP, the UN Evaluation Group and OECD-DAC has provided many lessons learned to further improve evaluative methodologies and practice. The present emphasis on providing evidence of effectiveness of humanitarian projects and programs provides a further strong stimulus to implement these lessons learned, to improve monitoring and evaluation performance and strive for increased contributions to evidence-based programming.

Notes

1 E.g. the OECD-DAC (1999), European Commission (2007), CARE International UK (2012), Global Fund (2012), ECB (2007), WFP (n.d.).

2 The most significant change (MSC) technique is a form of participatory monitoring and evaluation. The process involves the collection of significant change (SC) stories emanating from the field level, and the systematic selection of the most significant of these stories by panels of designated stakeholders or staff.

3 The ToR outline what is expected from and in an evaluation. It provides the basis for a contractual understanding between the commissioners of an evaluation and an evaluation team/consultant and it establishes the factors against which the success of the evaluation can be measured.

Conclusion

Liesbet Heyse, Andrej Zwitter, Rafael Wittek and Joost Herman

Humanitarian action often has to take place in highly complex environments, with strong pressures on coordination and cooperation between a heterogeneous group of stakeholders, tough budget constraints, demanding accountability requirements and considerable time pressure. These are highly adverse circumstances for any decision-maker. Yet it is these contexts within which aid organizations have to take informed decisions, among others when, where and how to intervene or with whom to cooperate. Not surprisingly, the humanitarian sector is often accused of having taken wrong or insufficiently informed decisions.

In the past decades, awareness has grown and urgency is felt to avoid wrong or uninformed decisions and resulting flaws in humanitarian aid provision. Collecting, evaluating and analyzing information and using this information for evidence-based programming is deemed key in preventing such repetitive mistakes. However, humanitarian crisis contexts are by far ideal settings for thorough and thoughtful information collection and analysis processes, if only due to contextual constraints such as safety risks, lack of (access to) information, doubtful reliability and validity of information, and the strategic use of information in humanitarian crises. Despite these constraints, we argue, the sector should strive for and can achieve increased impact by means of adopting an evidence-based programming approach.

A 'meta'-model for evidence-based humanitarian programming

Successful evidence-based programming requires robust and reliable tools for the collection and analysis of information on the context of the crisis and its stakeholders. We argue that such methods should meet a set of seemingly contradicting requirements. First, such tools should be both quick to apply and sufficiently thorough as to prevent superficial or simply wrong analysis. Second, such methods need to be quite generally applicable but also adaptable to the specifics of various crisis contexts (disasters, conflicts, complex emergencies, etc.) so that not only experts and specialists, but also a broad range of aid workers can perform these analyses. Third, in addition to an analytical framework, these methods should also provide guidelines for how to use the findings for programming purposes.

The humanitarian sector can already draw on many frameworks, tools and methods, which have proven useful in achieving more impact and preventing repetitive mistakes. We argue that these tools score well on some of the above-mentioned requirements but less on others. For example, some models (e.g. MIRA) can be labeled 'quick and coarse grained' whilst others, such as the PAR and the SLA frameworks, are 'slow but thorough'. In between these tools are political economy and arena approaches that provide a valuable actor-oriented lens, but offer less concrete entry points for humanitarian programming.

Our analysis of the pros and cons of existing models led us to develop a 'meta-model' for context analysis and evidence-based humanitarian programming: the Humanitarian Analysis and Intervention Framework (H-AID framework). This framework is supposed to complement existing approaches. We consider it to score high on both 'speed' and 'thoroughness': it can be applied by humanitarian practitioners working under time constraints, while yielding sufficiently reliable and valid information for decision-makers. Simultaneously, this framework aims to be generally applicable but locally adaptable as well as offering entry points for programming advice. The H-AID framework developed in this book is thus an attempt to help the sector in arriving at better 'informed decisions' when preparing and implementing humanitarian interventions.

Core components of the H-AID framework

As outlined in this book, the H-AID framework consists of three core components: Comprehensive Context Analysis, Intervention Analysis and Stakeholder Analysis (see Figure C.1). We briefly summarize them below.

Comprehensive Context Analysis: a multidisciplinary effort

CCA is firmly grounded in comprehensive security and securitization debates in International Relations. Five specific context dimensions crucial to CCA have been discussed: the political, economic, socio-cultural, food and health, and environmental contexts. These chapters demonstrate that a proper CCA of humanitarian crises is multidisciplinary in nature, and needs to combine insights from academic disciplines ranging from economics to medicine, and from anthropology to biology. The CCA also illustrates the power of visualizing the results. The 'radar chart' facilitates comparisons between different security dimensions, and identification of constraints and capabilities. These five context dimensions are always present and we therefore deem them crucial for any context analysis.

Intervention Analysis: doing good by doing nothing?

The H-AID framework also provides guidance for decision-makers, country and project managers, project proposal writers, assessment teams and humanitarian

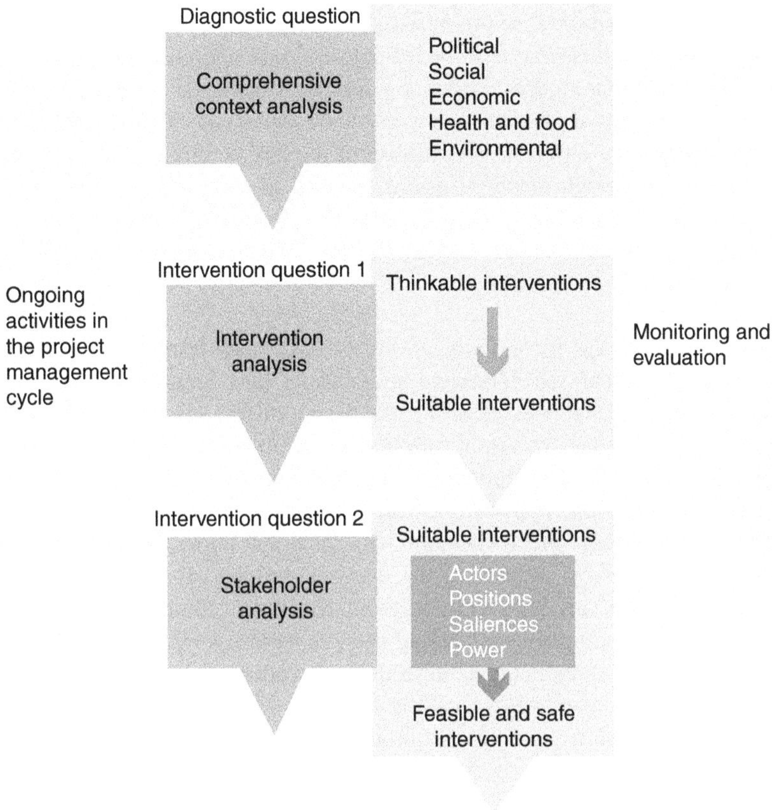

Figure C.1 Overview of the humanitarian analysis and intervention design framework.

experts to make the step from context analysis to intervention design. Using the method of theory-based *ex ante* evaluation allows to thoroughly think through, in advance, if action is required in a certain situation, and, if so, which action is then most suitable. Such an exercise consists of three steps: (1) identifying a set of interventions deemed suitable to a particular humanitarian situation; (2) outlining necessary context conditions for success of such identified interventions; (3) comparing these necessary context conditions to the actual context characteristics.

The H-AID framework purposefully follows a needs-based rather than a supply-driven approach in intervention analysis. This implies that an *ex ante* evaluation of proposed interventions might lead to the conclusion that these interventions might not be suitable or desirable, or might actually do more harm than good. Such an intervention analysis may help to prevent standardized responses and tunnel vision.

Stakeholder Analysis for negotiation and strategic action

The Stakeholder Analysis tool presented in this book helps to find out which of the suitable interventions may also be feasible and safe options. Stakeholder analyses in the humanitarian sector often are limited to an actor mapping exercise, which outlines a set of stakeholders present on the scene. The H-AID Stakeholder Analysis also requires identification of the key conflicting issues as they are related to the design of interventions, as well as the position, salience and power of these stakeholders. This information allows derivation of strategic insights for negotiation, both for purposes of implementing projects as well as of achieving organizational security.

Taken together, the three tools of the H-AID meta-framework (Comprehensive Context Analysis, Intervention Analysis and Stakeholder Analysis) provide information that is essential for the design of feasible interventions, and offer useful instruments for visualization. This information includes contextual capabilities and constraints, trends indicated by stakeholder interactions, stakeholder related risks and opportunities and the kinds of interventions that are appropriate. These three elements come with tools that allow for illustrating the information and thereby supporting the decision-making process. Finally, as a crucial component of any intervention design, this book provides guidelines and theoretical considerations for monitoring and evaluation as well as quality assurance. Each of these components requires that aid workers acquire a solid basis in information collection and analysis skills: they should know what information is valuable to collect, how to collect valid and reliable information, as well as to judge information quality, to analyze it, interpret it, scale and weigh it. We have also provided basic guidelines as how to achieve this.

Limitations to the H-AID framework

The H-AID framework has potential to assist the humanitarian sector in achieving evidence-based programming, but of course is not without limitations. We discuss four of these below.

Fuzzy categories and the risk of over- or underestimation

With the exception of the domains of food and health, there are usually no clearly defined sets of indicators or objective thresholds to assess the level of threat related to economic, social, political or environmental security. The lack of such objective criteria entails the risk of arbitrary judgments, and may lead to over- or underestimation of the threat to a specific domain. The analytical tools presented in the book are not able to solve this problem, and will not offer an 'objective' assessment of a situation. The tools do, however, offer guidelines for a comprehensive analysis, for identifiying 'known unknowns', and for being transparent about assumptions and choices made in the information collection and analysis process.

Interrelated domains and issues of causality

The five context dimensions are interrelated and influence each other. For example, threats to economic security may be the cause of threats to food security, which in turn can be a threat for social security. Though there are instruments to analyze and identify these interrelations – such as cause mapping, root cause or stream analysis techniques – the tools in this book do not provide much guidance as how to systematically take these interrelationships into account in terms of mapping and describing them, and drawing conclusions about implications for interventions. The radar graph is not a causal map, but a tool that offers a summary overview. Nevertheless, it may help to trigger questions about the repercussions of conditions and events in one context on threats to security in other contexts.

Stakeholder Analysis versus stakeholder dialogue

Stakeholder Analysis is a tool designed to maximize the outcomes of one single stakeholder. Though the tool may nevertheless be useful for resolving conflicts, it provides little guidance for how to solve collective action problems and achieve joint production across several stakeholders. For example, some crisis situations may require major collaborative efforts, which can only be achieved through inter-organizational coordination of the organizational field as a whole. The application of stakeholder analysis techniques that maximize the outcomes of single stakeholders (e.g. specific humanitarian organizations) may be less fruitful in such settings, unless flanked by other techniques directed towards finding sustainable joined solutions, like stakeholder dialogues.

Quantification and the risk of quasi-precision

Both CCA and stakeholder analysis use quantification as an analytical aid. While facilitating transparency about assumptions and decisions as well as fostering systematic comparison, one danger of this approach is that small errors may cumulate to large biases and that numbers are reified as a reality of their own. This may lead to an impression of quasi-precision. Consequently, such quantification steps should be done with caution. Interpretations should be grounded in thorough background knowledge on how the 'raw' information on context or stakeholder characteristics has been translated into numerical indicators, and in particular on the size of the potential error margins associated with the estimates. The H-AID framework therefore also shows the importance of a multi-method approach to humanitarian CCA, in which qualitative and quantitative information is used.

Adopting, routinizing and testing the H-AID framework

A key objective of this book is to show the added value of evidence-based programming and the use of tools such as the H-AID framework. A remaining challenge is how to build such a framework in existing organizational routines. This is not an easy task, as discussed in the chapter on monitoring, evaluation and learning in humanitarian organizations. One of the reasons is that the H-AID framework assumes an ongoing analytical and reflective process, closely aligned with monitoring processes in the project management cycle as well as internal management processes such as learning processes and security management. How is this achieved in aid organizations? Moreover, how can we make sure that the tool is not only used, but also used correctly and constantly improved?

It is beyond the scope of this book to provide a comprehensive analysis of how to build such sustainable routines (see also Chapter 13). We only sketch three organizational arrangements that might facilitate building such routines. First, the skills required for conducting a sound context, intervention and stakeholder analysis can be learned. Hence, *professionalization of training* of aid workers is an important safeguard to ensure a sound application of these techniques. These training efforts can be organized at the level of the humanitarian sector or within single humanitarian organizations. To ensure the quality of the training, cooperation between practitioners and academics has been proven to be very effective in the past, and should thus be sustained in the future. Even short training modules can be highly effective. Once acquired, these skills need to be practiced in order not to fade, so humanitarian organizations may benefit from making practical training and refreshing of these skills an integral part of their investments in its workforce.

Second, another ingredient for taking informed decisions is the *availability, quality and rapid accessibility of information*. The sector is to compliment for its efforts to improve the continuous and collective collection, systematization and accessibility of knowledge about emergency contexts, intervention processes and project performance. Great efforts to build systematic repositories have already been started, such as done in OCHA, ALNAP, ACAPs, various humanitarian organizations (CARE) and via internet applications such as the Humanitarian Genome search engine of lessons learnt in humanitarian aid (www.humaitariangenome.com).

Third, in order to remain useful, the tool itself needs to be constantly tested against real life applications, and subsequently adapted, expanded and improved. This requires continuous *feedback by practitioners* on the strengths and shortcomings of the tools as well as comparative case studies conducted by humanitarian studies students and scholars.

All in all, we consider the potential benefits of the H-AID framework in terms of speed, thoroughness, general applicability, local adaptability and generating programming advice to outweigh its limitations. It is an instrument that complements existing frameworks. We believe that its qualities have the potential to assist humanitarian organizations in their attempts to improve evidence-based programming. We hope that the framework will further prove its usefulness in future applications.

References

ACF Action Against Hunger International (2010) *Food security and livelihood assessments: a practical guide for field workers*, Available at: www.actionagainsthunger.org/publication/2010/04/food-security-and-livelihoods-assessments-practical-guide-field-workers [Accessed March 24, 2014].

Adato, M. (2001) *Combining qualitative and qualitative methods for program monitoring and evaluation: why are mixed designs best*, ALNAP Working Paper, Washington, DC: World Bank, Available at: https://openknowledge.worldbank.org/handle/10986/11063 [Accessed February 24, 2014].

Additional Protocol I to the Geneva Conventions relating to the Protection of Victims of International Armed Conflicts, Geneva (June 8, 1977).

Adler, P.A. and Adler, P. (2007) *Sociological odyssey: contemporary readings in introductory sociology*, Belmont, CA: Thomson Wadsworth.

Afek-Eitam, C. (forthcoming) *Organizational learning and evaluative practices in humanitarian response*, Dissertation, Groningen.

Alfaro, L., Chanda, A., Kalemli-Ozcan, S. and Syek, S. (2004) 'FDI and economic growth: the role of local financial markets', *Journal of International Economics* 64: 89–112.

Alina, C., Ciommi, M., Furia, D. and Odoardi, I. (2010) 'Reducing poverty as a driver of development', *Annals of the University of Oradea: Economic Science*: 208–13.

Allas, T. and Georgiades, N. (2001) 'New tools for negotiators', *McKinsey Quarterly* 2: 86–97.

ALNAP Active Learning Network for Accountability and Performance (2001) *Humanitarian action: learning from evaluation*, Annual Review 2001, London: ALNAP.

ALNAP Active Learning Network for Accountability and Performance (2003) *Participation by crisis-affected populations in humanitarian action: a handbook for practitioners*, London: ODI.

ALNAP Active Learning Network for Accountability and Performance (2008) *Rethinking the impact of humanitarian aid*, Background Paper for the 24th ALNAP Biannual, London: ALNAP, Available at: www.alnap.org/resource/5080 [Accessed February 25, 2014].

ALNAP Active Learning Network for Accountability and Performance (2012a) *The state of the humanitarian system 2012*, London: ODI.

ALNAP Active Learning Network for Accountability and Performance (2012b) *Evidence and knowledge in humanitarian action*, Background paper to the 28th ALNAP meeting, Washington, DC: ALNAP.

ALNAP Active Learning Network for Accountability and Performance (2013) *Evidence and knowledge in humanitarian action*, Background Paper for the 28th ALNAP Meeting, London: ALNAP.
Alons, S. (2009) *You cannot sweep the floor with just one lidi: the Javanese attitude to life and its importance for the perceived psychosocial well-being after the event of the 2006 in the Yogyakarta Province*, MA Thesis, Groningen: University of Groningen.
Amin, S., Quayes, S. and Rives, J.M. (2006) 'Market work and household work as deterrents to schooling in Bangladesh', *World Development* 34(7): 1271–86.
Anderson, M.B. (1999) *Do no harm: how aid can support peace – or war*, Boulder, CO: Lynne Rienner Publishers.
Anderson, M.B. and Woodrow, P.J. (1989) *Rising from the ashes: development strategies in times of disaster*, Boulder, CO: Westview Press/UNESCO.
Anderson, M.B. and Woodrow, P.J. (1990) *Disaster and development workshops: a manual for training in capacities and vulnerabilities analysis*, Boulder, CO: Westview Press.
Argyris, C. and Schön, D.A. (1974) *Theory in practice: increasing professional effectiveness*, San Fransisco, CA: Jossey-Bass.
Argyris, C. and Schön, D.A. (1978) *Organizational learning: a theory of action perspective*, Reading, MA: Addison-Wesley.
Astbury, B. and Leeuw, F.L. (2010) 'Unpacking black boxes: mechanisms and theory building in evaluation', *American Journal of Evaluation* 31(3): 363–80.
Austin, J.L. (1965) *How to do things with words*, Oxford: Clarendon Press.
Barnett, M. and Weiss, G.J. (eds) (2008) *Humanitarianism in question: power, politics and ethics*, Ithaca, NY/London: Cornell University Press.
Barry, N. and Barham, J. (2012) *Review of existing practices to ensure participation of disaster affected communities in humanitarian aid operations*, Report for the Directorate General for Humanitarian Aid (DG ECHO), Evaluation Sector.
Bauman, Z. (1991) *Thinking sociologically*, 2nd edn, Oxford: Blackwell.
Bazerman, M.H. and Neale, M. (1994) *Negotiating rationally*, New York: Simon & Schuster.
Beaudou, A. and Cambrézi, L. (1999) *Refugee camps and environment: landscape and deforestation in Dadaab region (Kenya)*, UNHCR, IRD, Available at: http://horizon.documentation.ird.fr/exl-doc/pleins_textes/doc34-05/010019833.pdf [Accessed April 11, 2014].
Bellegarde-Smith, P. (2011) 'A man-made disaster: the earthquake of January 12, 2010 – A Haitian perspective', *Journal of Black Studies* 42(2): 264–75.
de Benoist, B., McLean, E., Egli, I. and Cogswell, M. (eds) (2008) *Worldwide prevalence of anaemia 1993–2005: WHO global database on anaemia*, WHO, Available at: http://whqlibdoc.who.int/publications/2008/9789241596657_eng.pdf?ua=1 [Accessed March 24, 2014].
Bernard, H.R. (2011) *Research methods in anthropology*, Lanham, MD: Rowman, Altamira Press.
Bilton, T., Bonnett, K., Jones, P., Lawson, T., Skinner, D., Stanworth, M. and Webster, A. (2002) *Introductory sociology*, Basingstoke: Palgrave Macmillan.
Binder, A. (2013) 'Is the humanitarian failure in Haiti a system failure?' *International Development Policy/Revue internationale de politique de développement* 5(2), Available at: http://poldev.revues.org/1625 [Accessed October 22, 2013].
Bird, K., Booth, D. and Pratt, N. (2003) *Food security crisis in Southern Africa: the political background to policy failure poverty*, Public Policy Group, ODI, IDD, Birmingham University, Available at: www.odi.org.uk/sites/odi.org.uk/files/odi-assets/publications-opinion-files/2360.pdf [Accessed February 28, 2014].

Biswas, A.K. and Tortajada Quiroz, C. (1996) 'Environmental impacts of the Rwandan refugees on Zaire', *Ambio* 25(6): 403–8.

Black, R. (2001) *Environmental refugees: myth or reality? New issues in refugee research*, Working Paper no. 34, University of Sussex.

Blamey, A. and Mackenzie, M. (2007) 'Theories of change and realistic evaluation: peas in a pod or apples and oranges?' *Evaluation* 13(4): 439–55.

Blanchflower, D.G. and Oswald, A.J. (1994) *The wage curve*, Cambridge, MA/London: MIT Press.

den Boer, M. and de Wilde, J. (eds) (2008) *The viability of human security*, Amsterdam: Amsterdam University Press.

Bradt, D.A. (2009) *Evidence-based decision-making in humanitarian assistance*, HPN Network Paper no. 67, London: ODI, Available at: www.alnap.org/resource/7503.aspx [Accessed February 25, 2014].

Brown, C. (2007) 'The remaking of globalization', in: D. Held and A. McGrew (eds), *Globalization theory*, Cambridge: Polity Press.

Brown, L. (2011) *World on the edge: how to prevent environmental and economic collapse*, London: Earthscan.

Brown, L.D. and Moore, M.H. (2001) *Accountability, strategy, and international non-governmental organizations*, Working Paper no. 7 (April), Hauser Center for Non-Profit Organizations, Kennedy School of Government, Harvard University.

Brugha, R. and Varvasosvzky, Z. (2000) 'Stakeholder analysis: a review', *Health Policy and Planning* 15(3): 239–46.

Bryson, J.M. (2004) 'What to do when stakeholders matter', *Public Management Review* 6(1): 21–53.

Buchanan-Smith, M. and Cosgrave, J. (2013) *Evaluation of humanitarian action: pilot guide*, Available at: www.alnap.org/resource/8229 [Accessed November 14, 2013].

Buzan, B.G. (1991) *People, states and fear: an agenda for international security studies in the post-cold war era*, 2nd edn, Boulder, CO: Lynne Rienner.

Buzan, B.G., Waever, O. and de Wilde, J. (1998) *Security: a new framework for analysis*, Boulder, CO: Lynne Rienner.

Cahill, M. (ed.) (2007) *The pulse of humanitarian action*, New York: Ford University Press and Center for International Humanitarian Cooperation.

Cahuc, P. and Zylberberg, A. (2001) *Labor economics*, Cambridge, MA: Massachusetts Institute of Technology.

CARE International (2009) *Climate vulnerability and capacity analysis*, Available at: www.careclimatechange.org/cvca/CARE_CVCAHandbook.pdf [Accessed February 13, 2014].

CARE International (2012) *Guidance for designing, monitoring and evaluating peace building projects: using theories of change*, Available at: http://insights.careinternational.org.uk/media/k2/attachments/CARE_International_DME_for_Peacebuilding.pdf [Accessed February 25, 2014].

CARE and Save the Children (2010) *An independent joint evaluation of the Haiti earthquake humanitarian response*, Available at: www.canadahaitiaction.ca/sites/default/files/CARE,%20Save%20The%20Children%20report%20October%202010.pdf [Accessed April 18, 2014].

Carney, D. (2002) *Sustainable livelihoods approaches: progress and possibilities for change*, Toronto: DFID.

Carroll, R. and Phillips, T. (2010) 'Haiti aid effort "could have saved more"', *Guardian*, Available at: www.guardian.co.uk/world/2010/jan/22/haiti-aid-effort-earthquake [Accessed November 28, 2010].

Cartwright, N. (2012) 'Will this policy work for you? Predicting effectiveness better: how philosophy helps', *Philosophy of Science* 79(5): 973–89.
CDC Center for Disease Control and WFP World Food Programme (2005) *A manual: measuring and interpreting malnutrition and mortality*, Available at: www.unhcr.org/45f6abc92.html [Accessed February 28, 2014].
Cerny, P.G. (2010) *Rethinking world politics*, Oxford: Oxford University Press.
Chambers, R. and Conway, G.R. (1991) *Sustainable rural livelihoods: practical concepts for the 21st century*, IDS Discussion Paper no. 296, Brighton.
Chaudhuri, S. and Banerjee, D. (2010) 'FDI in agricultural land, welfare and unemployment in a developing economy', *Research in Economics* 64(4): 229–39.
Che, Y. and Wang, D.T. (2013) 'Multinationals, institutions and economic growth in China', *Asian Economic Journal* 27(1): 1–16.
Chiras, D.D. (2013) *Environmental science*, 9th edn, Burlington: Jones & Bartlett Learning.
Choudhury, E. and Ahmed, S. (2002) 'The shifting meaning of governance: public accountability of third sector organizations in an emergent global regime', *International Journal of Public Administration* 25(4): 561–88.
Christiansen, S.M. (2010) *Environmental refugees, a legal perspective*, Nijmegen: Wolf Legal Publishers.
Christopolos, I., Rodriguez, T., Schipper, L., Narvaes, E.A, Bayres, M., Karla, M., Buitrago, R., Gómez, L. and Pérez, F.J. (2009) *Learning from recovery after Hurricane Mitch: experience from Nicaragua, summary report*, Available at: http://preventionweb.net/go/12455 [Accessed July 10, 2010].
Clarke, P. and Ramalingam, B. (2008) *Organisational change in the humanitarian sector: ALNAP 7th review of humanitarian action*, Available at: www.alnap.org/resource/5231 [Accessed February 25, 2014].
Clayton, R.L., Sadeghi, A. and Talan, D.M. (2013) 'High-employment growth firms: defining and counting them', *Monthly Labor Review* 1: 3–13.
Cohen, C. and Werker, E.D. (2008) 'The political economy of natural disasters', *Journal of Conflict Resolution* 52(6): 795–819.
Cohen, R. and Kennedy, P.M. (2000) *Global sociology*, New York: New York University Press.
Collier, P. (2009) *Haiti: from natural catastrophe to economic security*, A report for the Secretary-General of the United Nations, Department of Economics, Oxford University.
Collier, P. and Hoeffler, A. (2004) 'Greed and grievance in civil war', *Oxford Economic Papers* 56(4): 563–95.
Collins English Dictionary (1994) 3rd edn, Glasgow: HarperCollins Publishers.
Collinson, S. (2002) *Politically informed humanitarian programming: using a political economy approach*, Humanitarian Practice Network paper no. 41, London: ODI.
Collinson, S. (ed.) (2003) *Power, livelihoods and conflict: case studies in political economy analysis for humanitarian action*, London: HPG.
Commission on Global Governance (1995) *Our global neighbourhood*, Oxford: Oxford University Press.
Connell, J.P., Kubisch, A.C., Schorr, L.B. and Weiss, C.H. (eds) (1995) *New approaches to evaluating community initiatives: concepts, methods and contexts*, Vol. 1, Washington, DC: Aspen Institute.
Convention I for the Amelioration of the Condition of the Wounded and Sick in Armed Forces in the Field, Arts 24–29, Geneva (August 12, 1949).

Convention II for the Amelioration of the Condition of Wounded, Sick and Shipwrecked Members of Armed Forces at Sea, Art. 36, Geneva (August 12, 1949).
Convention IV relative to the Protection of Civilian Persons in Time of War, Art. 20, Geneva (August 12, 1949).
Coolsaet, R. and van de Voorde T. (2006) *The evolution of terrorism in 2005: a statistical assessment*, Ghent: University of Ghent, Available at: https://biblio.ugent.be/input/download?func=downloadFile&recordOId=663241&fileOId=663242.
Coryn, C.L.S., Noakes, L.A., Westine, C.D. and Schröter, D.C. (2011) 'A systematic review of theory-driven evaluation practice from 1990 to 2009', *American Journal of Evaluation* 32(2): 199–226.
Coyne, C.J. (2013) *Doing bad by doing good: why humanitarian action fails*, San Francisco, CA: Stanford University Press.
Crescenzi, R. and Rodríguez-Pose, A. (2012) 'Infrastructure and regional growth in the European Union', *Papers in Regional Science* 91(3): 487–513.
Cristescu, A., Stanila, L. and Vasilescu, M.D. (2014) 'Earnings analysis at industry level', *Theoretical and Applied Economics* XXI(1): 27–36.
Crossan, M.M., Lane, H.W. and White, R.E. (1999) 'An organizational learning framework: from intuition to institution', *Academy of Management Review* 24(3): 522–37.
CSI Coping Strategies Index (2008) *The coping strategies index: field methods manual*, 2nd edn, CARE, Available at: www.fsnnetwork.org/sites/default/files/coping_strategies_tool.pdf [Accessed March 3, 2014].
Darcy, J. (2009) *Humanitarian diagnostics: the use of information and analysis in crisis response decisions*, Discussion paper prepared for the UN Food and Agricultural Organisation, London: ODI, HPG.
Darcy, J. and Hofmann, C.-A. (2003) *According to need? Needs assessment and decision making in the humanitarian sector*, London: ODI.
Darcy, J., Anderson, S. and Majid, N. (2007) *A review of the links between needs assessment and decision-making in response to food crises*, Study undertaken for the World Food Programme under the SENAC project, Rome: WFP.
Darcy, J. and Knox, P. (2013) *Evidence & knowledge in humanitarian action*, Background paper for the 28th ALNAP meeting, London: ALNAP, Available at: www.alnap.org/resource/7957 [Accessed March 9, 2014].
Darcy, J., Stobaugh, H., Walker, P. and Maxwell, M. (2013) *The use of evidence in humanitarian decision making*, ACAPS Operational Learning Paper, Boston, MA: Tufts University.
Davis, I., Haghebaert, B. and Peppiatt, D. (2004) *Social vulnerability and capacity analysis*, Discussion paper and workshop report, ProVention, Geneva.
Debevec, L. (2011) 'To share or not to share: hierarchy in the distribution of family meals in urban Burkina Faso', *Anthropological Notebooks* 17(3): 29–49.
DEC Disasters Emergency Committee (2014) *Haiti earthquake facts and figures*, Available at: www.dec.org.uk/haiti-earthquake-facts-and-figures [Accessed March 12, 2014].
Desrosiers, A. and St Fleurose, S. (2002) 'Treating Haitian patients: key cultural aspects', *American Journal of Psychotherapy* 56(4): 508–21.
de Deugd, N. (2005) *Ukraine and the issue of participation with the Euro-Atlantic security community*, Dissertation, Groningen.
DFID Department for International Development (1993) *Note on enhancing stakeholder participation in aid activities*, London: DFID.
DFID Department for International Development (1995) *Guidance note on how to do stakeholder analysis of aid projects and programmes*, London: DFID.

DFID Department for International Development (2001) *Sustainable livelihoods guidance sheets*, London: DFID.

Diamond, J. (2005) *Collapse: how societies choose to fail or succeed*, New York: Penguin Books.

Diamond, L.J. and Linze J.J. (1989) *Democracy in developing countries, vol 2: Africa*, Boulder, CO: Lynne Rienner.

van Dijkhorst, H. (2011) *Rural realities between crisis and normality: livelihood strategies in Angola, 1975–2008*, Dissertation, University of Wageningen.

Dijkzeul, D. and Herman, J. (2011) 'A matter of principles: humanitarian challenges', *Broker* 24, Available at: www.thebrokeronline.eu/Articles/A-matter-of-principles [Accessed June 16, 2014].

Donahue, A.K. and Tuohy, R.V. (2007) 'Lessons we don't learn: a study of the lessons of disasters, why we repeat them, and how we can learn them', *Homeland Security Affairs* 3(3).

Duarte, C.M., Lenton, T.M., Wadhams, P. and Wassmann, P. (2012) 'Abrupt climate change in the Arctic', *Nature Climate Change* 2: 60–2.

Ebrahim, A. (2005) 'Accountability myopia: losing sight of organizational learning', *Nonprofit and Voluntary Sector Quarterly* 34(1): 56–87.

ECB European Central Bank (2007) *Impact measurement and accountability in emergencies: the good enough guide*, Available at: www.ecbproject.org/downloads/resources/good-enough-guide-book-en.pdf [Accessed February 25, 2014].

ECHO European Commission for Humanitarian Aid (2005) *Manual project cycle management*, Available at: http://ec.europa.eu/echo/files/partners/humanitarian_aid/fpa/2003/guidelines/project_cycle_mngmt_en.pdf [Accessed April 18, 2014].

Edwards, M. and Hulme, D. (eds) (1996) *Non-governmental organisations: performance and accountability, beyond the magic bullet*, London: Earthscan Publications, Save the Children.

EIU Economist Intelligence Unit (2014a) 'Political instability index', Business Intelligence on 205 Countries, ViewsWire, Available at: http://viewswire.eiu.com/site_info.asp?info_name=social_unrest_table&page=noads [Accessed March 12, 2014].

EIU Economist Intelligence Unit (2014b) 'Political instability index: vulnerability to social and political unrest', Business Intelligence on 205 Countries, ViewsWire, Available at: http://viewswire.eiu.com/index.asp?layout=VWArticleVW3&article_id=874361472 [Accessed March 12, 2014].

Eriksen, H.T., Bal, E. and Salemink, O. (2010) *A world of insecurity: anthropological perspectives on human security*, London: Pluto Press.

ERM Environmental Resource Management (2007) 'Mainstreaming the environment in humanitarian response', Available at: https://www.humanitarianresponse.info/topics/environment [Accessed April 11, 2014].

European Commission (2007) *Evaluating humanitarian aid by and for NGOs*, Available at: http://ec.europa.eu/echo/files/evaluation/2007/humanitarian_guide.pdf [Accessed February 25, 2014].

Fan, L. (2013) *Disaster as opportunity? Building back better in Aceh, Myanmar and Haiti*, HPN Working Paper.

FAO Food and Agriculture Organization (2002) 'Glossary', in: *The state of food insecurity in the world 2001*, Rome, Available at: www.fao.org/docrep/003/y1500e/y1500e06.htm#P0_2 [Accessed April 7, 2014].

FAO Food and Agriculture Organization (2010) *Global forest resources assessment 2010*, FAO Forestry paper no. 163, Rome.

Fredrickson, B.L. and Kahneman, D. (1993) 'Duration neglect in retrospective evaluations of affective episodes', *Journal of Personality and Social Psychology* 65(1): 45–55.

Frerks, G. and Hilhorst, D. (2002) *New issues in refugee research: evaluation of humanitarian assistance in emergency situations*, Working Paper no. 56, UNHCR, Available at: www.unhcr.org/cgi-bin/texis/vtx/home/opendocPDFViewer.html?docid=3c8398434&query=%22New%20issues%20in%20Refugee%20Research%22 [Accessed February 25, 2014].

Gasper, D. (2000) 'Evaluating the "logical framework approach" towards learning-oriented development evaluation', *Public Administration and Development* 20(1): 17–28.

GHA Global Humanitarian Assistance (2007–2008) Annual report, Available at: www.globalhumanitarianassistance.org/wp-content/uploads/2010/07/2007-GHA-report.pdf [Accessed June 16, 2014].

Gibelman, M. and Gelman, S.R. (2001) 'Very public scandals: non-governmental organizations in trouble', *Nonprofit and Voluntary Sector Quarterly* 12(1): 49–66.

Global Fund (2012) *Monitoring and evaluation toolkit*, Available at: www.theglobalfund.org/en/mediacenter/announcements/2012-04-23_Release_of_the_fourth_edition_of_the_Monitoring_and_Evaluation_toolkit/ [Accessed June 16, 2014].

Gohmann, S.F. and Fernandez, J.M. (2013) 'Proprietorship and unemployment in the United States', *Journal of Business Venturing* 29: 289–309.

Goma Epidemiology Group (1995) 'Public health impact of Rwandan refugee crisis: what happened in Goma, Zaire, in July, 1994?', *Lancet* 345(8946): 339–44.

Greenhill, R. and Blackmore, S. (2002) 'Relief works', Jubilee Research at the New Economics Foundation.

Gross, R., Schoeneberger, H., Pfeifer, H. and Preuss, H.-J. (2000) *The four dimensions of food and nutrition security: definitions and concepts*, Available at: www.foodsec.org/DL/course/shortcourseFA/en/pdf/P-01_RG_Concept.pdf [Accessed March 3, 2014].

Groupe URD and Joint UNEP/OCHA Environment Unit (2013) Training course 'Integrating the environment into humanitarian action', in: Module 4 'The environment in the humanitarian programme cycle and in the project cycle', Geneva.

Gunaratnam, Y. (2003) *Researching 'race' and 'ethnicity': methods, knowledge and power*, London: Sage.

Gurr, T.R. (1974) *Why men rebel*, Princeton, NJ: Princeton University Press.

Haan, N., Majid, N. and Darcy, J. (2005) *A review of emergency food security assessment practice in Ethiopia*, London: ODI.

Hallam, A. (2011) *Harnessing the power of evaluation in humanitarian action: an initiative to improve understanding and use of evaluation*, ALNAP Working Paper, ALNAP, Available at: www.alnap.org/resource/6123 [Accessed June 16, 2014].

HAP Humanitarian Accountability Partnership (2008) *The 2008 humanitarian accountability report*, Available at: www.alnap.org/resource/8749 [Accessed February 25, 2014].

Harvey, P. and Bailey, S. (2011) *Cash transfer programming in emergencies*, Good Practice Review 11, London: ODI, HPN.

Heijmans, A. and Victoria, L.P. (2001) *Citizenry-based and development-oriented disaster response: experiences and practices in disaster management of the citizens' disaster response network in the Philippines*, Asian Disaster Preparedness Center.

Heintze, H.-J. and Zwitter, A. (eds) (2011) *International law and humanitarian assistance: a crosscut through legal issues pertaining to humanitarianism*, Berlin/Heidelberg: Springer Verlag.

Helman, C. (2001) *Culture, health and illness*, 4th edn, London: Arnold.

Heuer, R.J. and Pherson, R.H. (2010) *Structured analytic techniques for intelligence analysis*, Washington, DC: CQ Press.

Heyse, L. (2007) *Choosing the lesser evil: understanding decision making in humanitarian aid NGOs*, Aldershot: Ashgate Publishing Ltd.

Hibbs-Pherson, K. and Pherson, R.H. (2013) *Critical thinking for strategic intelligence*, Los Angeles, CA: CQ Press, Sage Publications.

Hilhorst, D. and Jansen, B.J. (2010) 'Humanitarian space as an arena: a perspective on the everyday politics of aid', *Development and Change* 41(6): 1117–39.

Hilhorst, D. and Serrano, M. (2010) 'The humanitarian space in Angola, 1975–2008', *Disasters* 34: S183–201.

Holt-Lunstad, J., Smith, T.B. and Layton, J.B. (2010) 'Social relationships and mortality risk: a meta-analytic review', *PLOS Medicine* 7(7), Available at: www.plosmedicine.org/article/info%3Adoi%2F10.1371%2Fjournal.pmed.1000316 [Accessed June 16, 2014].

Hoogerwerf, A. (1990) 'Reconstructing policy theory', *Evaluation and Program Planning* 13(3): 285–91.

Howard, M.C. (1989) *Contemporary cultural anthropology*, 3rd edn, New York: Harper/Collins.

HPN Humanitarian Practice Network/ODI Overseas Development Institute (2010) *Operational security management in violent environments*, HPN Good Practice Review 8, London.

IASC Steering Group on Security (2011) *Saving lives together: a framework for improving security arrangements among IGOs, NGOs and the UN in the field*, Available at: http://www.eisf.eu/resources/item/?d=1611 [Accessed April 4, 2014].

IASC Steering Group on Security (2012a) *Inter-agency real-time evaluation of the humanitarian response to the earthquake in Haiti 20 months after*, Available at: https://docs.unocha.org/sites/dms/Documents/IA%20RTE%20Haiti_phase%202%20final%20report.pdf [Accessed June 16, 2014].

IASC Steering Group on Security (2012b) *Multi-cluster/sector initial rapid assessment*, Available at: https://docs.unocha.org/sites/dms/CAP/mira_final_version2012.pdf [Accessed March 11, 2014].

ICRC International Committee of the Red Cross (2007a) *How to do a VCA: a practical step-by-step guide for Red Cross Red Crescent staff and volunteers*, Geneva: ICRC.

ICRC International Committee of the Red Cross (2007b) *VCA toolbox with reference sheets*, Geneva: ICRC.

ICRC International Committee of the Red Cross (2011) *Programme/project management: the results-based approach*, Geneva: ICRC.

ICRC International Committee of the Red Cross (2012) *Annual report*, Volume I, Geneva: ICRC.

IFRC International Federation of the Red Cross (2004) *World disasters report: focus on community resilience*, New York, Geneva: IFRC.

Inekwe, J.N. (2013) 'FDI, employment and economic growth in Nigeria', *African Development Review* 25(4): 421–33.

INSD Institute National de la Statistique et de la Démographie (2009) *Projections démographiques de 2007 a 2020, par région et province*, Ouagadougou: INSD.

International Review of the Red Cross (2011) 'Humanitarian debate: law, policy, action. The future of humanitarian action', *International Review of the Red Cross* 93(884) December.

IPCC Intergovernmental Panel on Climate Change (2013) *Fifth assessment report (AR5), summary for policymakers*, Available at: www.ipcc.ch/ [Accessed April 11, 2014].

Jisi, W. (2011) 'China's search for a grand strategy', *Foreign Affairs*, Available at: www.foreignaffairs.com/articles/67470/wang-jisi/chinas-search-for-a-grand-strategy [Accessed June 16, 2014].

Johns Hopkins and Red Cross/Red Crescent (2008) *Public health guide in emergencies*, 2nd edn, Geneva: IFRC.

Kahneman, D., Fredrickson, D.L., Schreiber, C.A. and Redelmeier, D.A. (1993) 'When more pain is preferred to less: adding a better end', *Psychological Science* 4(6): 401–5.

Keen, D. (2012) 'Aid and development in the context of conflict: some problems and pitfalls', at: High level expert forum on food insecurity in protracted crises, September 13–14, 2012, Rome, Available at: www.fao.org/fileadmin/templates/cfs_high_level_forum/documents/AID_AND_DEVELOPMENT_Keen_01.pdf [Accessed February 28, 2014].

Kelly, C. (2001) *Rapid environmental impact assessment: a framework for best practice in emergency response*, Disaster Management Working Paper 3, Benfield Greig Hazard Research Center, University College London, Available at: www.ucl.ac.uk/abuhc/resources/working_papers2 [Accessed April 11, 2014].

Kelly, C. (2004) 'Including the environment in humanitarian assistance', *Humanitarian Exchange* 27: 39–42.

Kelly, C. (2005a) *Guidelines for rapid environmental impact assessment in disasters, version 4.4*, Care International and Benfield Hazard Research Center, University College London, Available at: http://postconflict.unep.ch/humanitarianaction/documents/05_01-03.pdf [Accessed April 9, 2014].

Kelly, C. (2005b) *Emergency shelter environmental checklist: checklist-based guide to identifying critical environmental considerations in emergency shelter site selection, construction, management and decommissioning*, Care International and Benfield Hazard Research Center, University College London, Available at: https://www.humanitarianresponse.info/system/files/documents/files/Checklist%20based%20guide%20Emergency%20Shelter.pdf [Accessed April 11, 2014].

Kelly, C. (2008) *Damage, needs or rights? Defining what is required after disaster*, Disaster Management Working Paper no. 17, Aon Benfield Hazard Research Center, Available at: www.ucl.ac.uk/abuhc/resources/working_papers2 [Accessed April 11, 2014].

Kelly, C. (2013) *Mainstreaming environment into humanitarian interventions: a synopsis of key organisations, literature and experience*, Evidence on Demand, Available at: http://dx.doi.org/10.12774/eod_hd053.jul2013.kelly [Accessed April 11, 2014].

Kenny, C. (2009) *Why do people die in earthquakes? The costs, benefits and institutions of disaster risk reduction in developing countries*, Policy Research Working Paper no. 4823, World Bank Sustainable Development Network, Finance Economics and Urban Department.

Khouri-Padova, L. (2004) *Haiti: lessons learned*, Discussion Paper, UN Department of Peacekeeping Operations, Peacekeeping Best Practices Unit, New York, Available at: http://smallwarsjournal.com/documents/haitilessonslearned.pdf [Accessed June 16, 2014].

Kieffer, M. (1977) 'Disasters and coping mechanisms in Cakchiquel Guatemala: the cultural context', Intertect, Guatemala.

Klein, N. (2012) *Real wage, labor productivity, and employment trends in South Africa: a closer look*, International Monetary Fund Working Paper, 12/92.

Knowles, S. (1997) 'Which level of schooling has the greatest economic impact on output?' *Applied Economics Letters* 4(3): 177–81.
Krahenbuhl, P. (2011) *The militarization of aid and its perils*, ICRC Resource Centre, Available at: www.icrc.org/eng/resources/documents/article/editorial/humanitarians-danger-article-2011-02-01.htm [Accessed June 16, 2014].
Krantz, L. (2001) *The sustainable livelihood approach to poverty reduction: an introduction*, Sweden: SIDA.
Langen, C. (2007) *Toolkit sociologie*, Assen: Uitgeverij Van Gorcum.
Lasswell, H.D. and Kaplan, A. (1950) *Power and society: a framework for political inquiry*, New Haven, CT: Yale University Press.
Lautze, S. and Raven-Robert, A. (2006) 'Violence and complex humanitarian emergencies: implications for livelihoods models', *Disasters* 30(4): 383–401.
LeBillon, P. (2000) *The political economy of war: what relief agencies need to know*, Humanitarian Practice Network Paper no. 33, London: ODI.
Leeuw, F. (2003) 'Reconstructing program theories: methods available and problems to be solved', *American Journal of Evaluation* 24(1): 5–20.
Lehman J. (2009) *Environmental refugees: the construction of a crisis*, Prepared for the UHU-EHS Summer Academy 2009, Available at: www.ehs.unu.edu/file/get/4145 [Accessed April 11, 2014].
Lekha Sriram, C.L., King, J.C., Mertus, J.A., Martin Ortega, O. and Herman, J. (eds) (2009) *Surviving field research: working in violent and difficult situations*, Oxford: Routledge.
Lenton, T.M., Held, H., Kriegler, E., Hall, J.W., Lucht, W., Rahnstorf, S. and Schellnhuber, H.J. (2008) 'Tipping elements in the Earth's climate system', *Proceedings of the National Academy of Science of the United States of America* 105(6): 1786–93.
Levine, R. (1997) 'Financial development and economic growth: views and agenda', *Journal of Economic Literature* 35(2): 688–726.
Levine, S. and Chastre, C. (2004) *Missing the point: an analysis of food security interventions in the great lakes*, HPN Network Paper no. 47.
Levine, S., Crosskey, A. and Abdinoor, M. (2011) *System failure? Revisiting the problems of timely response to crises in the Horn of Africa*, HPN Network Paper no. 71, London: ODI, Available at: www.cmamforum.org/Pool/Resources/System-Failure-Horn-HPN-Levine-2011.pdf [Accessed March 28, 2013].
Liimatainen, H. and Pöllänen, M. (2013) 'The impact of sectoral economic development on the energy efficiency and CO2 emissions of road freight transport', *Transport Policy* 27: 150–7.
McGrew, A. (2007) 'Organised violence in the making (and remaking) of globalization', in: D. Held and A. McGrew (eds), *Globalization theory*, Cambridge: Polity Press.
McKinney, M.L., Schoch, R.M. and Yonavjak, L. (2013) *Environmental science: systems and solutions*, 5th edn, Burlington, MA: Jones & Bartlett Learning.
McKinnon, R.I. (1973) *Money and capital in economic development*, Washington, DC: Brookings Institution.
Maltais, A., Dow, K. and Persson, A. (2003) *Integrating perspectives on environmental security*, SEI Risk and Vulnerability Programme Report 2003-01, Available at: www.sei.se [Accessed April 11, 2014].
Malthus, T.R. (1798) *An essay on the principle of population*, London: J. Johnson, in St Paul's Church-yard.
March, J.G. (1994) *A primer on decision making: how decisions happen*, New York: Free Press.

Martell, L. (2010) *The sociology of globalisation*, Cambridge: Polity Press.

Maslow, A. (1943) 'A theory of human motivation', *Classics in the history of psychology*, Available at: http://psychclassics.yorku.ca/Maslow/motivation.htm [Accessed June 14, 2014].

Maxwell, D., Parker, J. and Stobaugh, H. (2013a) 'What drives program choice in food security crises? Examining the "response analysis" question', *World Development* 49: 68–79.

Maxwell, D., Sadler, L., Sim, A., Mutonyi, M., Egan, R. and Webster, M. (2008) *Emergency food security interventions*, Good Practice Review HPN no. 10.

Maxwell, D., Stobaugh, H., Parker, J. and McGlinchy, M. (2013b) *Response analysis and response choice in food security crises: a roadmap*, HPN Network Paper no. 73, London: ODI.

Medvedev, S. (2004) *Rethinking the national interest: Putin's turn in Russian foreign policy*, Marshall Papers 6, Garmisch-Partenkirchen.

Merriam-Webster Online Dictionary, Available at: www.merriam-webster.com [Accessed April 16, 2014].

Mertus, J. (2009) 'Maintenance of personal security: ethical and operational issues', in C.L. Sriram, J.C. King, J.A. Mertus, Martin-Ortega, O. and Herman, J. (eds) *Surviving field research: working in violent and difficult situations*, New York: Routledge.

Meyer, B.D. and Sullivan, J.X. (2012) 'Identifying the disadvantaged: official poverty, consumption poverty, and the new supplemental poverty measure', *Journal of Economic Perspectives* 26(3): 111–36.

Millennium Ecosystem Assessment (2003) *Ecosystems and human well-being: a framework for assessment*, Available at: www.millenniumassessment.org/en/Framework.html [Accessed April 11, 2014].

Millennium Ecosystem Assessment (2005) Available at: www.millenniumassessment.org [Accessed April 11, 2014].

de Milliano, C.W.J. (2012) *Powerful streams: exploring enabling factors for adolescent resilience to flooding*, PhD thesis, University of Groningen.

Mirwaldt, K., McMaster, I. and Bachtler, J. (2005) *Reconsidering cohesion policy: the contested debate on territorial cohesion*, Working Paper 08/5, EoRPA, University of Strathclyde.

Mitchell, R., Bradley, R. and Wood, D. (1997) 'Toward a theory of stakeholder identification and salience: defining the principle of who and what really counts', *Academy of Management Review* 22(4): 853–86.

Moneval Solutions (2013) *Project monitoring and evaluation: defining indicators*, Available at: www.moneval-solutions.com/defining-indicators/defining-indicators [Accessed December 9, 2013].

Morales, C. (2013) 'Alternatives to mainstreaming: grass roots perspectives on DRM', Conference paper, World Conference on Humanitarian Studies, Istanbul.

Morse, S., McNamara, N. and Acholo, M. (2009) *Sustainable livelihood approach: a critical analysis of theory and practice*, Geographical paper 289, University of Reading.

Mueller, J. (1990) *Retreat from doomsday: the obsolescence of major war*, New York: Basic Books.

Muggah, R. (2009) 'Haiti', in: OECD (ed.), *Bridging state capacity gaps in situations of fragility: lessons learned from Afghanistan, Haiti, South Sudan and Timor-Leste*, Paris: Partnership for Democratic Governance Experts Series, OECD.

Mulder, N. (1978) *Mysticism and everyday life in contemporary Java: cultural persistence and change*, Singapore: Singapore University Press.

Murray, H.A. (1938) *Explorations in personality*, New York: Oxford University Press.

Nagel, J. (1994) 'Constructing ethnicity: creating and recreating ethnic identity and culture', *Social Problems* 41(1): 152–76.

Newman, D.M. (2004) *Sociology: exploring the architecture of everyday life readings*, 5th edn, Thousand Oaks, CA: Pine Forge Press.

Nolan, P. and Lenski, G.E. (2006) *Human societies: an introduction to macrosociology*, 10th edn, Boulder, CO: Paradigm Publishers.

North, D.C. (1990) *Institutions, institutional change and economic performance*, Cambridge: Cambridge University Press.

North, D.C. (1992) 'Institutions, ideology and economic performance', *Cato Journal* 11(3): 477–88.

NRC Handelsblad (2013, October 14) 'Dit keer was India de cycloon voor', 10–11.

NRC Norwegian Refugee Council (2008) *Camp management toolkit*, Norwegian Refugee Council/Camp Management Project, Available at: www.nrc.no/arch/_img/9295458.pdf [Accessed April 15, 2014].

OCHA Office for the Coordination of Humanitarian Affairs (1999) *OCHA in 1999: activities and extrabudgetary funding requirements*, New York, Geneva, Available at: https://docs.unocha.org/sites/dms/Documents/ochain99.pdf [Accessed June 16, 2014].

OCHA Office for the Coordination of Humanitarian Affairs (2006) *OCHA in 2006: activities and extrabudgetary funding requirements*, New York, Geneva, Available at: https://docs.unocha.org/sites/dms/Documents/OCHAin2006.pdf [Accessed June 16, 2014].

OCHA Office for the Coordination of Humanitarian Affairs (2010) *Haiti revised humanitarian appeal*, Available at: http://ochaonline.un.org/humanitarianappeal/webpage.asp?MenuID=13737&Page=1843 [Accessed February 18, 2014].

Odhiambo, N.M. (2013) 'Inflation and economic growth in South-Africa: an empirical investigation', *Economics, Management, and Financial Markets* 4: 27–41.

ODI Overseas Development Institute (1999) *Planning tools: stakeholder analysis*, Available at: www.odi.org.uk/sites/odi.org.uk/files/odi-assets/publications-opinion-files/6459.pdf and www.foodsec.org/DL/course/shortcourseFK/en/pdf/trainerresources/PG_StakeHolder.pdf [Accessed June 16, 2014].

OECD-DAC Organisation for Economic Cooperation and Development-Development Assistance Committee (1999) *Guidance for evaluating humanitarian assistance in complex emergencies*, Available at: www.oecd.org/dac/evaluation/2667294.pdf [Accessed February 25, 2014].

OECD-DAC Organisation for Economic Cooperation and Development-Development Assistance Committee (2002) *Glossary of key terms in evaluation and results based management*, Available at: www.oecd.org/dataoecd/43/54/35336188.pdf [Accessed February 25, 2014].

Ogata, S. and Sen, A. (2003) *Human security now: final report*, New York: Commission on Human Security, Available at: www.unocha.org/humansecurity/chs/finalreport/English/FinalReport.pdf [Accessed April 7, 2014].

OHCHR Office of the High Commissioner for Human Rights (2006) 'Rule-of-law tools for post-conflict states mapping the justice sector', HR/PUB/06/2, New York/Geneva: Office of the United Nations High Commissioner for Human Rights, Available at: www.ohchr.org/Documents/Publications/RuleoflawMappingen.pdf [Accessed June 16, 2014].

Ohmae, K. (1995) *The end of the nation state and the rise of regional economics*, New York: Free Press.

Owen, T. (2002) 'Body count: rationale and methodologies for measuring human security', *Human Security Bulletin* 1(3), Available at: www.taylorowen.com/Articles/2002_%20Body%20Count.pdf [Accessed June 16, 2014].

Patrick, J. (2011) 'Haiti Earthquake response emerging evaluation lessons', *Evaluation Insights* 1(June), Department for International Development (DFID), Available at: www.oecd.org/countries/haiti/50313700.pdf [Accessed November 8, 2013].

Pawson, R. and Tilley, N. (2004) 'Realist evaluation', Paper prepared for the British Cabinet Office, Available at: www.communitymatters.com.au/RE_chapter.pdf [Accessed March 12, 2014].

Perlo-Freeman, S. and Webber, D.J. (2009) 'Basic needs, government debts and economic growth', *World Economy* 32(6): 965–94.

Perrin, B. (2012) *Linking monitoring and evaluation to impact evaluation: impact evaluation notes no. 2*, Available at: www.interaction.org/document/guidance-note-2-linking-monitoring-and-evaluation-impact-evaluation [Accessed November 13, 2013].

Physicians for human rights (2013) *Introduction to medical neutrality*, Available at: http://physiciansforhumanrights.org/library/other/introduction-to-medical-neutrality.html [Accessed June 16, 2014].

Pierre, A., Minn, P., Sterlin, C., Annoual, P.C., Jaimes, A., Raphaël, F., Raikhel, E., Whitley, R., Rousseau, C. and Kirmayer, L.J. (2010) 'Culture and mental health in Haiti: a literature review', *Santé Mentale en Haïti* 1: 13–42.

Proudlock, K., Ramalingam, B. and Sandison, P. (2009) *Improving humanitarian impact assessment: bridging theory and practice*, ALNAP 8th Review of Humanitarian Action: Performance, Impact and Innovation, London: ALNAP.

Provention Consortium (2007) *Tools for mainstreaming disaster risk reduction. vulnerability and capacity analysis*, Guidance Note 9.

Qayyum, U., Din, M. and Haider, A. (2014) 'Foreign aid, external debt and governance', *Economic Modelling* 37: 41–52.

Remington, T., Maroko, J., Walsh, S., Omanga, P. and Charles, E. (2002) 'Getting off the seeds-and-tools treadmill with CRS seed vouchers and fairs', *Disasters* 26(4): 316–28.

Rencoret, N., Stoddard, A., Haver, K., Taylor, G. and Harvey, P. (2010) *Haiti earthquake response context analysis*, London: ALNAP, UNEG.

Ride, A. and Bretherton, D. (2011) *Community resilience in natural disasters*, New York: Palgrave Macmillan, Available at: http://espace.library.uq.edu.au/view/UQ:268038 [Accessed June 28, 2013].

RIVM Rijksinstituut for Volksgezondheid en Milieu (2009) *The flash environmental assessment tool (FEAT)*, Available at: https://docs.unocha.org/sites/dms/Documents/FEAT_Version_1.1.pdf [Accessed April 11, 2014].

Roger, P.J. (2008) 'Using programme theory to evaluate complicated and complex aspects of interventions', *Evaluation* 14(1): 29–48.

Roslycki, L.L. (2011) *The soft side of dark power*, Dissertation, Groningen.

Sanders, C. (1997) 'Environment: where have all the flowers gone … and the trees … and the gorillas?', *Refugees Magazine* 110, Available at: www.unhcr.org/3b80c6e34.html [Accessed April 11, 2014].

Saracci, R. (1997) 'The World Health Organisation needs to reconsider its definition of health', *British Medical Journal* 314(7091): 1409–10.

Sarantakos, S. (2005) *Social research*, 3rd edn, New York: Palgrave Macmillan.

Sasketchewan Ministry of Education (2008) *Excerpts from review of evaluation frameworks*, Available at: www.idmbestpractices.ca/pdf/evaluation-frameworks-review.pdf [Accessed June 16, 2014].

Savage, G., Nix, T., Whitehead, C. and Blaire, J. (1991) 'Strategies for assessing and managing organizational stakeholders', *Executive* 5(2): 61–75.

Save the Children (2003) *Thin in the ground: questioning the evidence behind world bank-funded community nutrition projects in Bangladesh, Ethiopia and Uganda*, London: Save the Children UK.

Schmeer, K. (1999) *Guidelines for conducting a stakeholder analysis*, Health Reform Tool Series, Available at: www.who.int/management/partnerships/overall/GuidelinesConductingStakeholderAnalysis.pdf [Accessed June 16, 2014].

Schmeer, K. (2000) 'Stakeholder analysis guidelines', in: S. Scribner and D. Brinkerhoff (eds), *Policy toolkit for strengthening health sector reform*, Washington, DC: Health Sector Reform Initiative, Available at: http://info.worldbank.org/etools/docs/library/48545/RD1.PDF.pdf [Accessed June 14, 2014].

Sen, A.K. (1981) *Poverty and famines: an essay on entitlement and deprivation*, Oxford: Clarendon Press, Oxford University Press.

Shafik, N. (1994) 'Economic development and environmental quality: an econometric analysis', *Oxford Economic Papers* 46: 757–73.

Shapiro, I. (2006) 'Extending the framework of inquiry: theories of change in conflict interventions', in: D. Bloomfield, M. Fischer and B. Schmelzle (eds), *Social change and conflict transformation*, Dialogue Series No. 5, Berghof's Handbook of Conflict Transformation, Available at: www.berghof-handbook.net/dialogue-series/no.-5-social-change-and-conflict-transformation [Accessed January 10, 2014].

Sharpe, A., Arsenault, J.-F. and Harrison, P. (2008) *The relationship between productivity and real wage growth in Canada and OECD countries*, CSLS Research Reports 2008-08, Centre for the Study of Living Standards.

Shaw, E.S. (1973) *Financial deepening in economic development*, New York: Oxford University Press.

Smaje, C. (1997) 'Not just a social construct: theorising race and ethnicity', *Sociology* 31(2): 307–27.

Smedley, A. and Smedley, B.D. (2012) *Race in North America: the origin and evolution of a worldview*, Boulder, CO: Westview Press.

Smith, S.R. (1993) *Nonprofits for hire*, Cambridge, MA: Harvard University Press.

Sociology Guide (November 2013) Available at: www.sociologyguide.com [Accessed June 24, 2014].

Sommer, J.G. (1994) *Hope restored? Humanitarian aid to Somalia, 1990–1994*, Washington, DC: RPG.

Sperling, L. and McGuire, S.J. (2010) 'Persistent myths about emergency seed aid', *Food Policy* 35(3): 195–201.

Sphere Project (2011) The Sphere Handbook. *Humanitarian charter and minimum standards in humanitarian response*, 3rd edn, Rugby: Practical Action Publishing.

Steans, J. and Pettiford, L. (2001) *International relations: perspectives and themes*, Harlow: Pearson Education Limited.

Stein, D. and Valters, C. (2012) *Understanding of 'theory of change' in international development: a review of existing knowledge*, London: Asia Foundation/Justice and Security Research Programme.

Steuer, M.D. (2003) *The scientific study of society*, Boston, MA: Kluwer Academic Publishers.

Stewart, F. (2008) *Horizontal inequalities and conflict: understanding group violence in multiethnic societies*, London: Palgrave.

Stokman, F., van der Knoop, J. and van Oosten, R.C.H. (2013) 'Modeling collective decision-making', in: R. Wittek, T.A.B. Snijders and V. Nee (eds), *The handbook of rational choice social research*, Palo Alto, CA: Stanford University Press.

Strauss, J. and Thomas, D. (1998) 'Health nutrition and economic development', *Journal of Economic Literature* 36(2): 766–817.

Taleb, N.N. (2008) *The black swan*, London: Penguin.

TEC Tsunami Evaluation Coalition (2006) *Joint evaluation of the international response to the Indian Ocean tsunami: synthesis report*, ALNAP, Available at: www.alnap.org/ourwork/tec [Accessed October 8, 2013].

Tennant, D. and Abdulkadri, A. (2010) 'Empirical exercises in estimating the effects of different types of financial institutions' functioning on economic growth', *Applied Economics* 42(30): 3913–24.

The Concise Oxford Dictionary (1998) 9th edn, Oxford: Oxford University Press.

Thorsen, D. (2002) '"We help our husbands!" Negotiating the household budget in rural Burkina Faso', *Development and Change* 33(1): 129–46.

Tierney J. (2009, April 20) 'The richer-is-greener curve', *New York Times*.

Treasury Board of Canada (2012) *Theory-based approaches to evaluation: concepts and practices*.

Trouillot, M.-R. (1990) *State against nation: the origins and legacy of Duvalierism*, New York: Monthly Review Press.

Twigg, J. (2001) *Sustainable livelihoods and vulnerability to disaster*, Benfield Greig Hazard Research Center, Disaster Management Working Paper 2.

Twigg, J. (2004) *Disaster risk reduction. mitigation and preparedness in development and emergency programming*, Good Practice Review, London: HPN/ODI.

UN General Assembly (1948) *Universal Declaration of Human Rights*, G.A. Res. 217A (III), U.N. GAOR, 3rd Sess., U.N. Doc. A/810.

UNDP United Nations Development Programme Afghanistan Country Office (2008) 'State-building and government support programme', UNDP, Available at: www.undp.org.af/publications/KeyDocuments/Factsheets/sbgs/FactSheetSBGS-Dec08.pdf [Accessed June 16, 2014].

UNDP (1994) *Human development report 1994*, New York/Oxford: Oxford University Press.

UNDP (2009) *Post-disaster needs assessment, guidelines 2014*, Available at: www.recoveryplatform.org/ [Accessed April 15, 2014].

UNEP United Nations Environment Programme (2007) *Humanitarian action and the environment*, Joint info sheet with OCHA, Available at: http://postconflict.unep.ch/humanitarianaction/01_01.html [Accessed April 15, 2014].

UNEP United Nations Environment Programme (2008) *Environmental needs assessment in post-disaster situations: a practical guide for implementation*, Available at: https://www.humanitarianresponse.info/system/files/documents/files/UNEP_PDNA_draft.pdf [Accessed April 15, 2014].

UNEP United Nations Environment Programme (2011) *Mainstreaming the environment into humanitarian action*, Available at: https://www.humanitarianresponse.info/topics/environment/page/mainstreaming-environment-humanitarian-action [Accessed April 12, 2014].

UNEP United Nations Environment Programme (2012) *Integration of environmental issues in humanitarian programming*, Available at: https://www.humanitarianresponse.info/system/files/documents/files/Integration%20of%20environmental%20issues%20in%20humanitarian%20programming.pdf [Accessed April 12, 2014].

UNEP United Nations Environment Programme (2013) *Environment in humanitarian action*, Joint infosheet with OCHA, Available at: https://www.humanitarianresponse.info/system/files/documents/files/Environment%20into%20humanitarian%20action_leaflet_printingversion.pdf [Accessed April 15, 2014].

UNEP United Nations Environment Programme (2014a) *Disasters and conflicts, environmental mainstreaming*, Available at: www.unep.org/disastersandconflicts/CountryOperations/Sudan/Environmentalmainstreaming/tabid/54245/Default.aspx [Accessed April 14, 2014].

UNEP United Nations Environment Programme (2014b) *Environmental marker 2014*, Available at: www.unep.org/disastersandconflicts/Portals/155/countries/sudan/pdf/Environmental_Marker_Sector_Tip_Sheets_2014.pdf [Accessed April 15, 2014].

UNEP United Nations Environment Programme (2014c) *Environmental marker sector guidance*, Available at: www.unep.org/disastersandconflicts/Portals/155/countries/sudan/pdf/Environmental_Marker_2014_short_guidance.pdf [Accessed April 15, 2015].

UNHCR United Nations High Commissioner for Refugees (1990) *Assistance to refugees in Somalia: resolution/adopted by the General Assembly A/RES/45/154*, Available at: www.unhcr.org/3ae69ed80.html [Accessed April 11, 2014].

UNHCR United Nations High Commissioner for Refugees (1996) *Environmental guidelines*, Available at: www.refworld.org/pdfid/42a01c9d4.pdf [Accessed April 11, 2014].

UNHCR United Nations High Commissioner for Refugees (1997) *Social and economic impact of large refugee populations on host developing countries*, Available at: www.unhcr.org/print/3ae68d0e10.html [Accessed March 17, 2014].

UNHCR United Nations High Commissioner for Refugees (1998) *UNHCR global appeal 1999: the environment*, UNHCR Fundraising Reports, Available at: www.unhcr.org/print/3eff43e1d.html [Accessed March 12, 2014].

UNHCR United Nations High Commissioner for Refugees (2002) *Project planning in UNHCR: a practical guide on the use of objectives, outputs and indicators*, 2nd edn, UNHCR Division of Operational Support, Geneva, Available at: www.unhcr.org/3c4595a64.html [Accessed April 14, 2014].

UNHCR United Nations High Commissioner for Refugees (2005) *Environmental guidelines*, Available at: www.unhcr.org/3b03b2a04.html [Accessed April 11, 2014].

UNHCR United Nations High Commissioner for Refugees (2009a) *Environmental publications available from UNHCR*, Available at: www.unhcr.org/print/3b94c8364.html [Accessed April 9, 2014].

UNHCR United Nations High Commissioner for Refugees (2009b) *FRAME: framework for assessing, monitoring and evaluating the environment in refugee-related operations*, Available at: www.unhcr.org/4a97d1039.html [Accessed April 9, 2014].

US Government Accountability Office (1990) *Prospective evaluation methods: the prospective evaluation synthesis*.

Usón, A.U., Capilla, A.V., Bribián, I.Z., Scarpellini, S. and Llera Sastresa, E. (2011) 'Energy efficiency in transport and mobility from an eco-efficiency viewpoint', *Energy* 36(4): 1916–23.

Vogel, I. (2012) *Review of the use of 'theory of change' in international development*, Review report commissioned by the UK Department for International Development (DFID), London.

Volten, P.M.E. (ed.) (1992) *Bound to change: consolidating democracy in East Central Europe*, New York/Prague/Boulder, CO: Institute for EastWest Studies, Westwiew Press.

de Waal, A. (1997) *Famine crimes: politics and the disaster relief industry in Africa*, Oxford, Bloomington, IN: James Currey, Indiana University Press.

de Waal, A. (2010) 'The humanitarians' tragedy: escapable and inescapable cruelties', *Disasters*, Special Issue: the social dynamics of humanitarian action, 34(S2): S130–7.
WBZ-TV (2012) Available at: http://boston.cbslocal.com/2012/10/28/ [Accessed October 10, 2013].
Webb, P. and von Braun, J. (1994) *Famine and food security in Ethiopia: lessons for Africa*, London: John Wiley.
Weber, M. ([1922] 1978) *Economy and society*, Berkeley, CA: University of California Press.
Weiss, C. (1997) 'How can theory-based evaluation make greater headway?', *Evaluation Review* 21: 501–24.
Weiss, C. (2002) 'Which links in which theories shall we evaluate?' *New Directions for Evaluations* 87: 35–45.
WFP World Food Program (n.d.) *Monitoring and evaluation guidelines*, Available at: www.wfp.org/content/monitoring-and-evalutation-guidelines [Accessed February 25, 2014].
Whitman, J. and Pocock, D. (1996) *After Rwanda, the coordination of United Nations humanitarian assistance*, London/New York: Macmillan Press/St Martin's Press.
de Wilde, J. (2008) 'Environmental security deconstructed', in: H.G. Brauch, J. Grin, C. Mesjasz, P. Dunay, N. Chadha Behera, B. Chourou, P. Kameri-Mbote and P.H. Liotta (eds) *Globalisation and environmental challenges*, Hexagon Series on Human and Environmental Security and Peace, 3(1): 595–602.
WHO World Health Organization (n.d.) 'Public health', Available at: www.who.int/trade/glossary/story076/en/ [Accessed April 7, 2014].
WHO World Health Organization (1948) *Preamble to the constitution of the World Health Organization as adopted by the International Health Conference, New York, 19–22 June, 1946; signed on 22 July 1946 by the representatives of 61 states (official records of the World Health Organization, no. 2, p. 100) and entered into force on 7 April 1948*, Available at: http://whqlibdoc.who.int/hist/official_records/constitution.pdf [Accessed March 24, 2014].
WHO World Health Organization (1984) *Health promotion concepts and principles: a discussion document*, Regional Office for Europe, Copenhagen, Available at: http://whqlibdoc.who.int/euro/-1993/ICP_HSR_602__m01.pdf [Accessed March 24, 2014].
WHO World Health Organization (2011) 'Causes of death 2008: summary tables', Available at: http://apps.who.int/gho/data/node.main.890?lang=en [Accessed March 24, 2014].
WHO World Health Organization (2013) 'Diarrhoeal disease: fact sheet N°330', Available at: www.who.int/mediacentre/factsheets/fs330/en/ [Accessed March 24, 2014].
Wisner, B., Blaikie, P., Cannon, T. and Davis, I. (1994/2005) *At risk: natural hazards, people's vulnerability and disasters*, London/New York: Routledge.
Wisner, B., Gaillard, J.C. and Kelman, I. (2012) *The Routledge handbook of hazards and disaster risk reduction*, London/New York: Routledge.
Wolch, J.R. (1990) *The shadow state: government and voluntary sector in transition*, New York: The Foundation Center.
World Bank (2004) *Monitoring and evaluation (M&E): some tools, methods and approaches*, Available at: www.worldbank.org/ieg/ecd/tools/ [Accessed November 14, 2013].
World Bank (2012) *Haiti earthquake PDNA (post-disaster needs assessment): assessment of damage, losses, general and sectoral needs*, Annex to the Action Plan for National Recovery and Development of Haiti, Washington, DC: World Bank, Available at: http://documents.worldbank.org/curated/en/2010/03/16394500/haiti-earthquake-pdna-post-disaster-needs-assessment-assessment-damage-losses-general-sectoral-needs [Accessed June 16, 2014].

Wright, R.T. and Boorse, D.F. (2011) *Environmental science: toward a sustainable future*, 11th edn, Upper Saddle River, NJ: Pearson.

WWF World Wildlife Fund, Inc. and American Red Cross (2010) *Green recovery & reconstruction: training toolkit for humanitarian aid*, Available at: http://green-recovery.org [Accessed April 15, 2014].

Zhu, X., van Ommeren, J. and Rietveld, P. (2009) 'Indirect benefits of infrastructure improvement in the case of an imperfect labor market', *Transportation Research Part B* 43: 57–72.

Zwitter, A. and Lamont, C.K. (2014) 'Enforcing aid in Myanmar: state responsibility and humanitarian aid provision', in A. Zwitter, C.K. Lamont, H.-J. Heintze and J. Herman (eds), *Humanitarian action: global, regional and domestic legal responses*, Cambridge: Cambridge University Press.

Index

Page numbers in *italics* denote tables, those in **bold** denote figures.

4i/5i frameworks for organizational learning 184–7, **186**
9/11 terrorist attacks 32, 41

Abdulkadri, A. 77
absolute scaling 37
abstraction and prognosis 39–40, 41
accountability 3, 173, 176, 181, 182
Aceh, Indonesia 64, 65, 90
Active Learning Network for Accountability and Performance (ALNAP) 3, 5, 172, 173, 187, 194
actor analysis 10, 13, 18–19
Afghanistan 63
after action reviews *179*; *see also* monitoring and evaluation
aggregation and weighting of data 46–7, 54; economic context 79–85; food context 109, *112–13*; health context *110–11*; political context 66–9; socio-cultural context *96–7*, 98, *98*
agriculture: Foreign Direct Investment (FDI) 76; intensification 107; irrigation 119–20; state interventions 107
aid provision *see* mechanisms of aid provision
Alina, C. 71, 72
ALNAP *see* Active Learning Network for Accountability and Performance (ALNAP)
analysis, levels of 34–5, **37**
Anderson, Mary 19–20
anemia 106, 108
applicability of frameworks 15, 25–7, *26*; Capabilities and Vulnerabilities Analysis (CVA) 20; Multi-Cluster/Sector Initial Rapid Assessment (MIRA) 18; Political Economy/Arena Approach (PolEc) 19; Pressures and Release model (PAR) 23; Sustainable Livelihoods Framework (SLF) 24
appreciative enquiries *179*, 180
Argyris, C. 187
avalanches *123*

Bam earthquake, Iran 91
Bangladesh 140–1
bargaining 166–7
bargaining chips 167, **167**, 168
baseline data 58–61, *59*, **60**, **61**
baseline studies *178*, 180
basic goods, accessibility of 72, *80*, 83
basic needs fulfillment 78
beneficiaries, including in monitoring and evaluation 174, 181
biases 54
Blackmore, S. 78
Blaikie, P. 21
Body Mass Index (BMI) 108, 109, *112–13*
breast-feeding 108
Bretherton, D. 86, 89, 90
Brown, Lester 119–20, 121, 123–4
budget deficits 78
Bureau of Labor Statistics (BLS) 73
Burkina Faso 92
Business Employment Dynamics (BED) program 73
bystanders 165, **166**

Cahuc, P. 71
CAME approach 130
Camp Management Toolkit 128
Capabilities and Vulnerabilities Analysis (CVA) 6, 12, **13**, 19–21, 25–7, **25**, *26*

capacity-building 65
carbon dioxide emissions 120
Cartwright, N. 140–1
case fatality rate (CFR) 105
cash transfers 138–9
cause analysis 13
CCA *see* Comprehensive Context Analysis (CCA)
Centers for Disease Control and Prevention (CDC), USA 106
Central America 95
Centre for Research on the Epidemiology of Disasters (CRED) 106
Che, Y. 76
Chernobyl disaster, Ukraine 122, 124
childbirth 105
children: child labor 78; measurements 108, 109; mortality 99
China 31
cholera 104
Christoplos, Ian 180
climate change 40, 41, 104, 120–1
Cluster approach 3
CMO *see* Context-Mechanism-Outcomes (CMO) approach
coalitions 164
Commission on Global Governance 32–3
common courtyard (*lakou*) system, Haiti 91
community environmental action planning (CEAP) 129
comparison and prognosis 40–1
Comprehensive Context Analysis (CCA) 9, *9*, 190; dimensions and levels of analysis 33–7, **37**; measurement of context dimensions 53–6; measurement scale 46; pre-post-incident analysis 58–61, *59*, **60**, **61**; threat perception and securitization 37–41; visualization and interpretation 56–8, **57**, *57*, *58*; *see also* data collection; economic context; environmental context; food context; health context; political context; socio-cultural context
Comprehensive Security 30
compromise position 159–60, *159*, **160**
confidence intervals 51, 55–6, *56*; radar graphs 56–7
conflicts 64; and food security 102; and health 101
consumer protection 77
consumption-based poverty measures 72
context analysis 13
context dimensions 33–4, 35–7, **37**, 53–6; confidence intervals 55–6, *56*; pre-post-incident analysis 58–61, *59*, **60**, **61**; scaling of 54–5, *54*, *58*; *see also* economic context; environmental context; food context; health context; political context; socio-cultural context
Context-Mechanism-Outcomes (CMO) approach 135–6, **136**; areas for intervention 141; CMO configurations 146–8, **147**; context conditions 139–41, 145–6, **145**; mechanisms 136–9, 143–5, **144**; suitable intervention options 142–8; thinkable intervention options 141–2, **143**
contractual institutions 77, *82*, 84
Convention on the Rights of Persons with Disabilities 101
Coping Strategies Index (CSI) 109
corporate control 77
counterinsurgency strategies 63
Crossan, Marie 184–5
crude mortality rate (CMR) 104–5
cultural beliefs 88
culture *see* socio-cultural context
currency 74–5, *81*, 83
CVA *see* Capabilities and Vulnerabilities Analysis (CVA)

data analysis, obstacles to 6–7; statistical 182
data collection 182–4, *183*; aggregation and weighting 46–7; confidence intervals 51; decomposition 44–5; obstacles to 6–7; omnibus context 43–4; operationalization 45; quantification 46; quantitative score cards 129–30; reliability of analysis 14–15; reliability of data 45, 47–51, *49*; semi-structured interviews 184; story-telling techniques 184; surveys 182; triangulation 48–51, 184; trustworthy and accurate sources 47–8, *49*, *50*; validity of analysis 14–15; verification techniques for data sources 48, *49*, *50*
de Waal, Alex 102
deal breakers 166–7, **167**
debt relief 78
decision-making: operational 176; strategic 176
decomposition of data 44–5
deforestation 116, 119, 121, 122, 125
democratization, third wave of 31
depositary institutions 77, *82*, 84
development programs 91

diarrheal diseases 99, 105
disaster relief programs 91
disaster risk 21, **22**
Disaster Risk Management, Central America 95
domestic violence 39
double loop learning 187
drinking water 99, 104, 127
duration neglect 41
dynamic pressures 21, **22**, 23

early warning mechanisms *178*, 180, *183*
Earth system 116–17
earthquakes 122, *123*; Bam, Iran 91; Haiti 1–2, 5, 68–9, 87; Indian Ocean 2, 64, 65, 181; Pakistan 89; *see also* Fukushima disaster, Japan
easy wins 167, **167**
economic context 70–85; aggregation and weighting of data 79–85; emergency seed aid interventions **145**, 146; and food security 107; indicators 79, *80–1*, *82*; local dimensions 71–5, *80–1*, 83; operationalization of data 79; regional and national dimensions 75–9, *82*, 83–4
Economist Intelligence Unit (EIU) 66, *66*
ecosystem services 117, **118**, 121, 122
ecosystems 115–17
education 77–8, *82*, 84; Haiti 93
effective power of stakeholders 157, *159*, 160, 161, **162**, 169
El-Hinnawi, Essam 123
Emergency Capacity Building project (ECB) 174
emergency food interventions: context 139–41; information sources 142; mechanisms 138–9; thinkable and suitable interventions **143**
Emergency Nutrition Network 108
emergency seed aid interventions: CMO configurations 147–8, **147**; context conditions 145–6, **145**; mechanisms 143–4, **144**
Emergency Shelter Environmental Checklist 128
employment 72–3, 74, *80*, 83
employment protection 71
energy efficiency 75, *81*, 83
entitlements theory 102
entrepreneurship 72, 74, 83
environment: defined 115; markers 130
environmental assessments 127–8; hazardous chemicals 130; refugee and returnee situations 129–30
environmental context 115–30; ecosystem services 117, **118**, 121, 122; ecosystems 115–17; emergency seed aid interventions 145–6, **145**; environmental and natural disasters 122–4, *123*; environmental refugees 123–5, **124**; environmental resources 117–21; environmental security 119; and food security 107; and health security 104; humanitarian action and environmental management 125–30; indicators 126–7
environmental degradation, refugee camps 116, 125
environmental impact assessments (EIA) 128
Environmental Kuznets Curve (EKC) 75
Environmental Needs Assessment 128
environmental refugees 123–5, **124**
Epicentre-Paris 106
European Council 71
evaluation *see* monitoring and evaluation
evidence-based programming 4–6, 132–4
ex ante evaluation *see* theory-based *ex ante* evaluation
external mechanisms of aid provision 90–1
external monitoring and evaluation 180–1; *see also* monitoring and evaluation
extreme weather events 120

Failed States Index 65
famine 101, 102
feasible interventions 134; *see also* Stakeholder Analysis
final evaluations *179*, *183*
financial institutions 76–7, *82*, 84
financial legislation 77, *82*, 84
Flash Environmental Assessment Tool (FEAT) 130
flooding 121, 122, *123*
focus group discussions 184
food aid 102
Food and Agriculture Organization (FAO) 101, 108
food context 99–109; aggregation and weighting of data 109, *112–13*; and economics 107; emergency seed aid interventions 145, **145**; and environment 107; food and nutrition security 101–2, *112–13*; indicators 108; operationalization of data 106–9; and political security 107–8; *see also* health context
food insecurity: acute 101, 102; chronic 101, 102

food interventions *see* emergency food interventions
foreign assistance 91
foreign debt 78, *82*, 84
Foreign Direct Investment (FDI) 76, *82*, 84
forests: deforestation 116, 119, 121, 122; water regulation 122
formal social structures 90
formative final evaluations *179*, *183*
fossil fuels 120
fragile livelihoods and unsafe locations 21, **22**
Framework for Assessing, Monitoring and Evaluating the Environment in Refugee-related Operations (FRAME) 129–30
free food distributions 138–9, 140
Fukushima disaster, Japan 64, 68
Fund for Peace 65

gender: socialization 88; stratification 92
George Mason University, USA 66
Global Ecosystem 116–17
Global Hunger Index (GHI) 109
Global Outbreak Alert and Response Network (GOARN) 106
Global War on Terrorism 32
global warming 120–1; *see also* climate change
globalization 94–5
goals-based monitoring and evaluation 181–2; *see also* monitoring and evaluation
goals-free monitoring and evaluation 182; *see also* monitoring and evaluation
Goma, Zaire 105
good governance 77
Good Humanitarian Donorship Initiative 3
Good Practice Review 13, 19, 27
Green Recovery and Reconstruction Toolkit 128
green revolution 102
Greenhill, R. 78
Greenland Ice Sheet 120–1

H-AID *see* Humanitarian Analysis and Intervention Design Framework (H-AID)
Haiti: political environment 68; security situation 94; social change 95; social control 94; social identity 89; social stratification 92–3; social structures 91
Haiti earthquake 1–2, 5, 68–9, 87
hazards 21, **22**
health context 99–109; aggregation and weighting of data *110–11*; emergency seed aid interventions 145, **145**, 146; and environment 104, 127; health security and public health 100–1; indicators 104–6; levels of health security *110–11*; operationalization of data 103–6; *see also* food context
health services, indicators 105
Heavily Indebted Poor Countries (HIPC) initiative 78
Heijmans, A. 20
Heintze, H.-J. 64
Helsinki Memorandum of Understanding 64, 65
hidden unemployment 72–3, *80*
high-growth businesses 73–4
Hilhorst, D. 18, 19
Human Development Reports 33–4
human rights 32, 100, 101
human security 30, 31, 33–4, 35–6, 40
Humanitarian Accountability Project (HAP) 174, 181
humanitarian aid workers, security 4
Humanitarian Analysis and Intervention Design Framework (H-AID) 12, 27–8; adopting, routinizing and testing 193–4; limitations 192–3; overview 7–10, **8**, *9*, 190–2, **191**; *see also* Comprehensive Context Analysis (CCA); Intervention Analysis; Stakeholder Analysis
humanitarian crisis analysis frameworks, existing 12–14, **13**; balancing core qualities of 14–16, **16**; Capabilities and Vulnerabilities Analysis (CVA) 6, 12, **13**, 19–21, 25–7, **25**, *26*; comparing and combining 25–8, **25**, *26*; Multi-Cluster/Sector Initial Rapid Assessment (MIRA) 12, **13**, 16–18, 25–7, **25**, *26*; Political Economy/Arena Approach (PolEc) 12, **13**, 18–19, 25–7, **25**, *26*; Pressures and Release model (PAR) 6, 12, **13**, 21–3, **22**, 25–7, **25**, *26*; Sustainable Livelihoods Framework (SLF) 12, **13**, 23–4, **24**, 25–7, **25**, *26*
Humanitarian Genome 194
Humanitarian Ombudsman Project 173–4
hurricanes 64, 120, 122, 180

IASC Steering Group on Security 3
ice sheet melting 120–1
IMF 78
impact assessments *179*, *183*
implementation theory 135
incidence rates 105

income 71, 73, *80*
income-based poverty measures 71, 72
independent monitoring and evaluation 180–1
India 122–3, 140–1
Indian Ocean earthquake and tsunami 2, 64, 65, 181
indicators: action/output 175; direct measurement 182; economic context 79, *80–1*, *82*; environmental context 126–7; expert opinion of 79; food context 108; health context 104–6; impact 175; input 175; lack of defined set 192; monitoring and evaluation 175; outcome 175; political context 65–6, *66*, *67*, *68*; predefined lists of 45, 54; process 175; proxy-indicators 55; quantitative and qualitative 54–5
Indonesia: Aceh 64, 65, 90; Java 88; Javanese cultural beliefs 88
Infant Mortality Rate 105
infectious diseases 99, 104
inflation rates 74–5, *81*
influencers 165, **166**
informal social structures 90
information collection and analysis: obstacles to 6–7; *see also* data collection
information repositories 194
innovation capacity 75, *81*, 83
Interagency Standing Committee (IASC) *see* Multi-Cluster/Sector Initial Rapid Assessment (MIRA)
Intergovernmental Panel on Climate Change (IPCC) 120
International Covenant on Economic, Social and Cultural Rights (ICESCR) 101
International Humanitarian Law 101
international interference 30–1
International Red Cross and Red Crescent Movement 63, 91, 181
interval scales 46
Intervention Analysis 6, 9–10, *9*, 132–4, 190–1; *see also* theory-based *ex ante* evaluation
interventions: thinkable 133, 141–2, **143**
intervention theory 135
investment institutions 77, *82*, 84
Iran 91
iron deficiencies 106, 108
irrigation 119–20
issues: bargaining chips 167, **167**, 168; deal breakers 166–7, **167**; defining 152–5; easy wins 167, **167**

Japan 64, 68
justice sector programs 65

Kelly, Charles 126, 128–9
Kieffer, M. 90
known unknowns 43–4, 192
Krahenbuhl, Pierre 63

labor adjustment costs 71
lakou (common courtyard) system, Haiti 91
large businesses 76, *82*, 84
learning, organizational 184–7
learning processes 140
levels of analysis 34–5, **37**
Levine, R. 76–7
liberalization 77
life expectancy 105
Liimatainen, H. 75
literacy 77–8, *82*, 84
livelihood assets 23, **24**
livelihood models 6; *see also* Sustainable Livelihoods Framework (SLF)
livelihood strategies 24, **24**
lobbying 167
loneliness 86
low birth weight 105

M&E *see* monitoring and evaluation
macro forces 21, **22**
malaria 99, 104
malnutrition 99, 101, 102, 109; emergency food interventions 138–41
Maltais, A. 119
Malthus, Thomas 102
marginalization 21, **22**
Maternal Mortality Rate (MMR) 105
measurement scales 46; scaling of context dimensions 54–5, *54*; nominal scales 46; ordinal scales 46
mechanisms of aid provision 136–9; external 90–1; identifying 143–5, **144**; internal 90
media opinion 173, 174
mental illnesses 106
Mertus, J. 66
Meyer, B.D. 72
micronutrient deficiencies 101, 106, 108
Mid-Upper Arm Circumference (MUAC) measurements 108, 109, *112–13*
MIRA *see* Multi-Cluster/Sector Initial Rapid Assessment (MIRA)
monetization, degree of 74, *81*, 83

monitoring and evaluation 171–88; 4i/5i frameworks for organizational learning 184–7, **186**; definitions and purposes 174–81, **177**; evolution of 172–4; internal 180; meta-evaluations *179*, 180, *183*; methodologies and tools 181–4, *183*; mid-term *178*, *183*; performance and context monitoring *178*, 183; qualitative methods 182–4, *183*; real time *178*, 180, *183*; self- 180; strategies *9*; summative final evaluations *179*, *183*; theory-based 182; types 177, *178–9*, 180–1
monocultures 107
moral hazard 76
mortality rates 104–5
MUAC *see* Mid-Upper Arm Circumference (MUAC) measurements
mudslides *123*
Multi-Cluster/Sector Initial Rapid Assessment (MIRA) 12, **13**, 16–18, 25–7, **25**, *26*
multinationals 76
Myanmar 88, 93, *123*

national budgets 78, *82*, 84
national food production indexes 108
National Institute for Public Health and the Environment (RIVM), Netherlands 130
natural disasters 122–4, *123*; applicability of frameworks 23; tropical cyclones 64, 88, 120, 122–3, *123*, 180; *see also* earthquakes
natural resources 117–19; depletion 119–20, 121; lack of access to 21, **22**; pollution 120–1
needs assessments *178*, 180, *183*, 184
needs-based analysis 133–4
negotiation strategies 164–5, 166–7, **166**
Newman, D.M. 86, 87, 88, 90, 94–5
non-communicable diseases (NCDs) 99, 104
non-state actors, and security 31–2
nuclear holocaust 39–40
nutrition response analysis tool 6
nutrition security *see* food context

Odhiambo, N.M. 75
OECD-DAC 174, 181, 187
Ogata, S. 100
oil spills 122
omnibus context 43–4
O'Neill, M. 63
Operational Security Management 3
operationalization of data 45; economic context 79; food context 106–9; health context 103–6; socio-cultural context 95
organizational learning 184–7
organizational security 6, 12, 13, 64
overfishing 119
over-pumping 119–20
Owen, T. 40

Pakistan 89, 121, 122
Palme, Olof 30
Pan American Health Organization (PAHO) 106
PAR *see* Pressures and Release model (PAR)
Patrick, J. 1–2, 5
Peak-End rule 41
performance standards 174
Perlo-Freeman, S. 72, 77–8
Philippines 20
Pierre, A. 89, 91, 92–3, 95
political context 63–9; aggregation and weighting of data 66–9; defined 65; emergency seed aid interventions **145**, 146; and food security 107–8; history 64–5; indicators 65–6, *66*, *67*, *68*
Political Economy/Arena Approach (PolEc) 12, **13**, 18–19, 25–7, **25**, *26*
Political Instability Index 66, *66*
political risk, defined 65
political security: defined 65; and food security 107–8
Pöllänen, M. 75
pollution 120–1
Post Disaster Needs Assessment (PDNA) 128; Haiti 68
potential followers 164, **166**
poverty 71–2, 75, *80*, 83
precautionary measures, natural disasters 122–3
pregnancy 105, 108, 109
pre-post-incident analysis 58–61, *59*, **60**, **61**
Pressures and Release model (PAR) 6, 12, **13**, 21–3, **22**, 25–7, **25**, *26*
problem analysis 5–6
program theory 135, 143–4, 182
programming advice 15, 25–7, *26*; Capabilities and Vulnerabilities Analysis (CVA) 20–1; Multi-Cluster/Sector Initial Rapid Assessment (MIRA) 18; Political Economy/Arena Approach (PolEc) 19; Pressures and Release model (PAR) 23; Sustainable Livelihoods Framework (SLF) 24

progression of vulnerability model 21–3, **22**
project and program management cycles 2–3, 176–7, **177**
property rights protection 76
public health 100; *see also* health context
public opinion 173, 174

qualitative data 54–5
qualitative score cards 129–30
quality assurance 173–4; *see also* monitoring and evaluation
quality of analysis 14–15
quantification: of data 46; limitations 193; in Stakeholder Analysis 169
quantitative data 54–5
quantitative monitoring and evaluation methods 182–4, *183*
quantitative score cards 129–30
quick scan model 79–83

race, Haiti 92–3
radar graphs 56–8, **57**, 141, 193
radiation levels 124
randomized controlled trials (RCTs) 5
Rapid Environmental Impact Assessment in Disasters (REA) 128–9
Realist Evaluation tradition 135
recession 72
refugee camps 116, 125
refugees: environmental 123–5, **124**; impact on environment 116, 125, 129–30; Rwandan 105
relative deprivation 40–1
reliability: of analysis 14–15; of data 45, 47–51, *49*
religion: Haiti 89; Java 88
remittances, Haiti 95
Rencoret, N. 94
rent-seeking 76
resource allocation 78–9, *82*, 84
resources: lack of access to 21, **22**; *see also* natural resources
response analysis 6
Ride, A. 86, 89, 90
risk assessment 2
risk diversification 75, 77
risk management 2, 3
risks to interventions 147–8, **147**
RIVM *see* National Institute for Public Health and the Environment (RIVM), Netherlands
root causes 21, **22**, 23
rule of law programs 65
Russia 31
Rwanda genocide 173
Rwandan refugees, Goma, Zaire 105

salience of stakeholders 156–7, *158*, *159*, 164–5, 166–7, **166**, **167**, 169
sanctions, social control 93–4
sanitation 99, 104
Save the Children 63
savings mobilization 75, 77
scales *see* measurement scales
Schön, D.A. 187
score cards 129–30
sea level rise 121, 124
sea-ice melting 120–1
securitization 37–41
security: human 30, 31, 33–4, 35–6, 40; humanitarian aid workers 4; new concepts of 30–3; organizational 6, 12, 13, 64; state 35–6; threat perception 37–41; traditional concept of 29–30; *see also* human security
security levels 54–5, *54*, 57, *57*, *58*
security risk management 3
security sector training 65
security situation, Haiti 94
seed aid interventions *see* emergency seed aid interventions
self-employment 72, 74
self-monitoring and evaluation 180
semi-structured interviews 184
Sen, Amartya 100, 102
service utilization records 182
sex-for-aid scandals, West Africa 173
sexual violence 39; Haiti 94
shapers 164–5, **166**
Shapiro, I. 137
single loop learning 187
SLF *see* Sustainable Livelihoods Framework (SLF)
small businesses 73–4, *80*, 83
social change 94–5, *97*
social class systems 92
social control 93–4, *97*
social environment 35–6
social identity 87–9, *96*
social inequality 92
social institutions 90
social organizations 90
social stratification 92–3, *97*
social structures 89–91, *96*
socialization 88
societal deficiencies 21, **22**
socio-cultural context 86–98; aggregation

and weighting of data *96–7*, 98, *98*; emergency seed aid interventions **145**, 146; operationalization of data 95; social change 94–5, *97*; social control 93–4, *97*; social identity 87–9, *96*; social stratification 92–3, *97*; social structures 89–91, *96*
Somalia 31
Southern Africa, food insecurity 107
sovereignty 30–1
speed and thoroughness of analysis 14–15, 25–7, *26*; Capabilities and Vulnerabilities Analysis (CVA) 20; Multi-Cluster/Sector Initial Rapid Assessment (MIRA) 17–18; Political Economy/Arena Approach (PolEc) 19; Pressures and Release model (PAR) 23; Sustainable Livelihoods Framework (SLF) 24
Sphere Handbook 105, 115
Sphere Project 3, 106, 174
stability analysis 160–2, **162**
Stakeholder Analysis *9*, 10, 134, 149–70, 192; expert information 150–1; interpretation 169–70; limitations 193; measurement scale 46; position of stakeholders 156, *158*, *159*; power of stakeholders 157, *158*, *159*, 164–5, **166**, 169; quantification 169; step 1: issue definition 152–5; step 2: stakeholder identification 155–6; step 3: stakeholder description 156–7, *158*, *159*; step 4: outcome continuum 157–60, **160**; step 5: stability analysis 160–2, **162**; step 6: stakeholder classification 162–4, *163*; step 7: negotiation landscape 164–5, **166**; step 8: relationship analysis 166–9, **167**
stakeholders: bystanders 165, **166**; effective power 157, *159*, 160, 161, **162**, 169; influencers 165, **166**; position 156, *158*, *159*; potential followers 164, **166**; power 157, *158*, *159*, 164–5, **166**, 169; professional opinion of indicators 79; salience 156–7, *158*, *159*, 164–5, 166–7, **166**, **167**, 169; shapers 164–5, **166**
star-bursting 43–4
state security 35–6
state-building 65
statistical analysis 182
story-telling techniques 184
street violence 39
structural approach to humanitarian crises 12–13
suitable interventions 133–4, 142–8, **143**
summative final evaluations *179*, *183*
supply-driven responses 4
surveys 182
sustainable development 75
Sustainable Livelihoods Framework (SLF) 12, **13**, 23–4, **24**, 25–7, **25**, *26*

targeted food distributions, problems with 1
Tennant, D. 77
terrorist attacks 32, 38–9, 41
theoretical compromise position 159–60, *159*, **160**
theory of change 135
theory-based *ex ante* evaluation 9–10
theory-based *ex ante* evaluation 133–6, **136**, 191; areas for intervention 141; CMO configurations 146–8, **147**; context conditions 139–41, 145–6, **145**; mechanisms 136–9, 143–5, **144**; suitable intervention options 142–8, **143**; thinkable intervention options 141–2, **143**
theory-based monitoring and evaluation 182
thinkable interventions 133, 141–2, **143**
threats: acute *54*, 55; latent *54*, 55; manifest *54*, 55; objective security 39; overestimation of 192; perception of security threats 37–41; perception 37–41; underestimation of 192
toxic waste 124
trading 74, 83
traffic accidents 38, 39, 41
training 194; security sector 65
transforming structures and processes 24, **24**
transparency 3, 173
transport infrastructure 73, *80*, 83
triangulation 48–51, 184
triple loop learning 187
tropical cyclones 64, 88, 120, 122–3, *123*, 180
Tsunami Evaluation Coalition 181
tsunamis 122; *see also* Fukushima disaster, Japan; Indian Ocean earthquake and tsunami
typhoons 120, 122

Ukraine 31, 122, 124
Under Five Mortality Rate (U5MR) 105
undernourishment 101
unemployment 72–3, 74, *80*

United Nations 3, 30, 101; civilian police (UNPOL) 94; environmental management 130; Evaluation Group 187; Stabilization Mission in Haiti (MINUSTAH) 94
United Nations Development Programme (UNDP) 33–4
United Nations Environment Programme (UNEP) 123, 125, 126, 130
United Nations High Commissioner for Refugees (UNHCR) 123, 125, 129
United States 31; 9/11 terrorist attacks 32, 41; hurricanes 64, 120, 180
unknown unknowns 43–4
unsafe conditions 21, **22**
Usón, A.U. 73

validity of analysis 14–15
verification techniques for data sources 48, *49*, *50*
Victoria, L.P. 20
vitamin deficiencies 101, 106
volcanic eruptions 122, *123*
von Braun, J. 101
vulnerability: and environmental disasters 122–3; progression of vulnerability model 21–3, **22**; *see also* Capabilities and Vulnerabilities Analysis (CVA)
vulnerability context 24, **24**

wages 71, 73, *80*
Wang, D.T. 76
war 64; and food security 102; and health 101
water: drinking 99, 104, 127; over-pumping 119–20; regulation by forests 122
water, sanitation and hygiene (WASH) 99, 104
water tables, falling 119–20, 124
Webb, P. 101
Webber, D.J. 72, 77–8
weighting of data *see* aggregation and weighting of data
welfare 72, 73, 78
Wisner, B. 21
Woodrow, P. 19–20
World Bank 68, 108, 140–1
World Food Programme (WFP) 102
World Health Organization (WHO) 100, 106, 109
World Summit on Social Development, Copenhagen 71
Worldwatch Institute 123

Zwitter, A. 64
Zylberberg, A. 71

For Product Safety Concerns and Information please contact our EU
representative GPSR@taylorandfrancis.com
Taylor & Francis Verlag GmbH, Kaufingerstraße 24, 80331 München, Germany

www.ingramcontent.com/pod-product-compliance
Ingram Content Group UK Ltd.
Pitfield, Milton Keynes, MK11 3LW, UK
UKHW022131190926
475702UK00013B/1751
9781138236622